T0271271

Principles of Astrophysical Fluid Dynamics

Fluid dynamical forces drive most of the fundamental processes in the Universe and so play a crucial role in our understanding of astrophysics. This comprehensive textbook introduces the fluid dynamics necessary to understand a wide range of astronomical phenomena, from stellar structures to supernovae blast waves, to accretion discs.

The authors' approach is to introduce and derive the fundamental equations, supplemented by text that conveys a more intuitive understanding of the subject, and to emphasise the observable phenomena that rely on fluid dynamical processes. It has been developed for use by final year undergraduate and starting graduate students of astrophysics, based on the authors' many years of teaching their astrophysical fluid dynamics course at the University of Cambridge. The book contains over 50 exercises.

CATHIE CLARKE is Reader in Theoretical Astrophysics at the University of Cambridge and Director of Studies in Astrophysics at Clare College. She developed the original course in astrophysical fluid dynamics as part of Part II Astrophysics in 1996 and delivered the course 1996–9. Her research is based on accretion disc theory and star formation (both of which are strongly based on fluid dynamics). She has taught extensively within the University of Cambridge, having also delivered lecture courses in statistical physics, mathematical methods and galactic dynamics, and has supervised for a variety of courses within the Natural Sciences and Mathematics Triposes.

BOB CARSWELL is Professor of Astronomy at the University of Cambridge. He lectured the Part II Astrophysics course on astrophysical fluid dynamics 2000–3, and developed the course notes to reflect a revised syllabus to include accretion discs and some MHD concepts. He has also given courses in relativity to both third-year and fourth-year undergraduates, as well as specialist courses on gaseous nebulae at the postgraduate level. His research relates to quasars, the intergalactic medium, and large-scale structure.

Principles of Astrophysical Fluid Dynamics

**Cathie Clarke and
Bob Carswell**
University of Cambridge

CAMBRIDGE
UNIVERSITY PRESS

University Printing House, Cambridge CB2 8BS, United Kingdom

Published in the United States of America by Cambridge University Press, New York

Cambridge University Press is part of the University of Cambridge.

It furthers the University's mission by disseminating knowledge in the pursuit of education, learning and research at the highest international levels of excellence.

www.cambridge.org
Information on this title: www.cambridge.org/9781107666917

First published 2007
First paperback edition (with corrections) 2014

A catalogue record for this publication is available from the British Library

ISBN 978-0-521-85331-6 Hardback
ISBN 978-1-107-66691-7 Paperback

Contents

Preface

The material in this book is based on lecture notes of a course on astrophysical fluid dynamics which has been given for several years to third-year students at the University of Cambridge. There are several excellent books which cover fluid dynamics from a terrestrial standpoint, but very few provide a full introduction to the concepts and methods used to deal with the highly compressible flows which arise in astrophysical contexts. Our aim with this book is to provide just such an introduction, and we hope that it will also serve as a reference volume for advanced undergraduate and graduate students.

Several people have provided input at various stages of the preparation of this book. In particular we thank Jim Pringle, Donald Lynden-Bell and Giuseppe Lodato for their help. We are also grateful to the students who have taken the course at Cambridge for correcting typographical errors in the lecture notes, drawing our attention to parts where the description was less clear than it should have been, and helping us to develop the exercises.

Chapter 1
Introduction to concepts

Stated most simply, fluids are 'things that flow'. This definition distinguishes between liquids and gases (both fluids) and solids, where the atoms are held more or less rigidly in some form of lattice. Of course, it is always possible to think of substances whose status is ambiguous in this regard, such as those, normally regarded as solids, which exhibit 'creep' over sufficiently long timescales (glass would fall into this category). Such borderline cases do not undermine the fact that the vast majority of substances can be readily classified as fluid or not. If they *are* fluids, then it is important to understand the general problem of how they flow, and under what circumstances they attain equilibrium (i.e. do not flow). These issues, in an astronomical context, form the subject of this book.

There is also a more subtle point about the sorts of systems that can be described as fluids. Although fluids are always in practice composed of particles at a microscopic level, the equations of hydrodynamics treat the fluid as a continuous medium with well-defined macroscopic properties (e.g. pressure or density) at each point. Such a description therefore presupposes that we are dealing with such large numbers of particles locally that it is meaningful to average their properties rather than following individual particle trajectories. In a similar vein, we may also, for example, treat the dynamics of stars in galaxies as a form of fluid dynamical problem: in this case the 'particles' are stars rather than atoms or molecules but the same principles may be used to determine the mean properties of the stars (such as velocity or density) in each region.

In this book, however, we will mainly be concerned with conventional fluids, i.e. liquids and gases. In fact, since the liquid state is hardly encountered apart from in the high pressure environments of planetary surfaces and interiors, our focus will very much be on the gas phase (although some of these gases, such as the degenerate gases that compose neutron stars and white dwarfs, bear little resemblance at a microscopic level to conventional gases under laboratory conditions). However, the key property of all gases, as opposed to liquids, is that they are far more *compressible*. Although in many terrestrial applications involving subsonic flows, even gases behave approximately incompressibly, this is not the case in astronomical contexts where flows are frequently accelerated (often by gravity) to high Mach number. This book is therefore not able to make the simplifying assumption, often introduced at an early stage of standard texts on terrestrial fluid mechanics, that the flow is incompressible. Likewise we cannot assume that the battery of techniques for the solution of incompressible flows can be simply generalised to the present case.

1.1 Fluids in the Universe

The baryonic matter in the Galaxy (i.e. conventional matter composed of protons and neutrons) is divided between stars and distributed gas roughly in the ratio 5:1. For the Universe as a whole the ratio is uncertain, but the gas fraction is considerably higher.

Stars are gaseous bodies (mainly hydrogen and helium) with temperatures that range between millions of kelvin in their centres, where nuclear reactions occur, and thousands of degrees at the surface. An easily remembered property of the Sun is that its *mean* density is the same as that of water, but this statistic does not convey its strong internal density stratification (the density at the centre exceeds that at the photosphere – visible surface of the Sun – by 11 orders of magnitude). For some purposes, the interior of stars may be regarded as static, i.e. in a state of force balance between gravity and outwardly directed pressure gradients. In practice, the gas in many stars is subject to internal motions such as convection currents and low amplitude internal oscillations (acoustic modes, see Figure 1.1). Above the photosphere, the gas density falls with increasing height, and the temperature rises, attaining $30\,000\,K$ in the so-called chromospheric region where many stellar emission lines originate. At larger distances still, the gas may be magnetically heated to temperatures of around $10^6\,K$, this coronal region being a strong source of X-rays. We however caution that the low densities in these latter regions mean that a fluid dynamical treatment is not necessarily appropriate (see Section 1.2).

The other main fluid component in the Universe, the distributed gas in the interstellar medium (ISM) and intergalactic medium (IGM), is much more diverse in its properties. For example, the mean density

Fig. 1.1. A cut-away illustration showing a spherical harmonic mode of oscillation for acoustic waves in the Sun. (Illustration from Global Oscillation Network Group/National Solar Observatory/AURA/NSF)

of gas in the Milky Way is the easily remembered 1 particle per cubic centimetre, or a million per cubic metre (extraordinarily dilute compared with 2.7×10^{25} particles per cubic metre of gas at standard terrestrial pressure and temperature). This figure however averages over a rich multi-phase medium, comprising warm atomic gas (at $\sim 10^4$ K), a hot phase (at 10^6 K) heated mainly by supernova explosions and a cold molecular phase, which may be as cool as 10 K if well shielded from radiation from bright stars. The density contrasts between these phases are extreme, from ~ 1000 particles per cubic metre for the hot phase to 10^5–10^6 particles per cubic metre for the warm, atomic phase to $\sim 10^8$ particles per cubic metre as a mean for molecular clouds; the densest cores within these clouds have densities in excess of 10^{13} particles per cubic metre. Outside galaxies the densities can be considerably lower, with large regions containing $\lesssim 1$ particle per cubic metre.

Although stars, the ISM and IGM together constitute the bulk of the fluids in the Universe, there are a number of other examples of fluids of astrophysical interest. These include stellar winds, jets and accretion discs on a wide range of scales. Nor should it be forgotten that an important category of stars – the white dwarfs and neutron stars – are also fluid, though with an equation of state – relation between pressure, density and temperature – that is quite different from conventional gases under laboratory conditions. Similarly, the internal structure of the giant planets may be determined as a fluid dynamical problem, although here there are considerable uncertainties surrounding the correct equation of state for hydrogen under extreme conditions of density and pressure in the relevant temperature range.

The above thumbnail sketch has stressed the diversity of fluids in the Universe and has perhaps suggested that their study will involve consideration of much complex microphysics. One of the beauties of fluid dynamics is however the way that the microphysical complexity is all contained in a single parameter – the equation of state. Once armed with an equation of state, the fluid dynamicist can simply insert this into the general equations of fluid dynamics. This is not to suggest that the outcome of this exercise is necessarily simple, but there is a pleasing generality that runs through the subject. We will, for example, be able to deduce the structure of white dwarfs and neutron stars as readily as of stars composed of conventional gas, simply because they each have well-defined equations of state and in each case we can consider the stars to be in a state of equilibrium. However, this discussion anticipates much of the contents of later chapters. Before we can embark on deriving the fluid equations, it is necessary that we now introduce some important fluid dynamical concepts.

1.2 The concept of a 'fluid element'

The fluid equations are based on the concept of a *'fluid element'*, i.e. a region over which we can define local variables (such as density, temperature, etc.). The size of the region is assumed to be such that:

(i) It is small enough that we can ignore systematic variations across it for any variable we are interested in, i.e. the region size ℓ_{region} is much smaller than a scale length for change of any relevant variable q (where a scale length is the scale over which q varies by order unity). So

$$\ell_{\text{region}} \ll \ell_{\text{scale}} \sim \frac{q}{|\boldsymbol{\nabla} q|}. \tag{1.1}$$

(ii) It is large enough that the element contains sufficient particles for us to ignore fluctuations due to the finite number of particles ('discreteness noise'). Thus if the number density of particles per unit volume is n, we require that

$$n\ell_{\text{region}}^3 \gg 1. \tag{1.2}$$

The above two criteria must be met by *any* system that is describable as a fluid. In addition, *collisional* fluids must satisfy the following criterion:

(iii) The fluid element is large enough that the constituent particles 'know' about local conditions through colliding with each other, so if the mean free path is λ, we require that

$$\ell_{\text{region}} \gg \lambda. \tag{1.3}$$

What do we mean by a collisional fluid? The essential point is that if the particles in a fluid interact with each other (which does not necessarily imply that they physically collide), then they will attain a distribution of, say, particle speed that maximises the entropy of the system at that temperature. (For this discussion we can simply regard the entropy of a system as being a measure of the number of microscopically distinct configurations that correspond to a given macroscopic property of the fluid locally.) Therefore a collisional fluid at a given temperature and density will have a well-defined distribution of particle speeds in the local rest frame, and hence a corresponding pressure. Thus one can derive an equation of state (relationship between density, temperature and pressure) for a collisional fluid. Almost all the fluids we consider in this book, be they ideal gases or degenerate, are collisional fluids with corresponding equations of state. However, as noted above, we can also consider systems of orbiting stars in a galaxy, or grains in Saturn's rings, as being fluids, even though the particles now do not interact sufficiently frequently to satisfy criterion (1.3) above. In this case, the distribution of particle speeds locally does not correspond to an entropy maximum but instead depends on the initial distribution of particle speeds. The fact that one cannot write down an equation of state for collisionless systems means that it is a hard problem, whose solution is non-unique, to find, for example, the structure of stellar orbits in a galaxy in equilibrium. In this book we will avoid this difficulty by considering collisional fluids (i.e. conventional liquids and gases).

It should be stressed that a fluid element is purely a conceptual entity – ℓ_{region} does not enter into the fluid equations. However, if a system is to be described by the fluid equations, its properties must be such that there are values of ℓ_{region} that simultaneously satisfy the conditions above.

1.3 Formulation of the fluid equations

There are two (sensible) approaches to formulating the equations for mass density, momentum and energy in a fluid:

(i) Eulerian description
Consider a small volume at a fixed spatial position. The fluid flows through the volume with physical variables specified as functions of time and the (fixed) position of the small volume: density $\rho = \rho(\mathbf{r}, t)$,

temperature $T = T(\mathbf{r}, t)$, etc. The change of any measurable quantity as a function of time at that position is $\partial/\partial t$ of the quantity, evaluated at the *fixed position*.

(ii) Lagrangian description
In this approach one chooses a particular fluid element and examines the change in variables (density, temperature, etc.) in that particular element. So the (spatial) reference system is comoving with the fluid. Thus one might examine the behaviour of $\rho = \rho(\mathbf{a}, t)$, where $\mathbf{a}$ is a label for a particular fluid element – which might be the coordinates at a chosen initial time, for example. The time derivative is now a partial one at a fixed $\mathbf{a}$ (i.e. for a fixed bit of the fluid), and the rate of change with respect to time for a *fixed element* is denoted D/Dt.

In the Lagrangian description, position is not an independent variable but instead $\mathbf{r} = \mathbf{r}(\mathbf{a}, t)$. The Lagrangian description refers to the world as seen by an observer riding on a fluid element (e.g. a river viewed from a boat adrift on it); the Eulerian one refers to the world as seen at a fixed spatial position (e.g. a river viewed from the bank).

The Lagrangian approach is useful if the path of an individual element is important, e.g. when a particular element has some property which distinguishes it from all the others. Usually this is not the case; however, one can think of particular instances where it is important in astronomy (e.g. tracing the trajectory of a parcel of gas that has been enriched by heavy elements as it is ejected into the interstellar medium by a supernova). The Eulerian approach is usually more useful if the motion of individual fluid elements is not of interest. It is particularly useful for *steady* flows, i.e. those where the quantities at a given position do not change. Then $\partial/\partial t$ of all variables $= 0$ everywhere. Steady flows have no special properties in Lagrangian descriptions since in a steady flow an individual element still changes its properties in general as it goes from place to place.

This conceptual split between the Eulerian and Lagrangian formulations translates into two entirely distinct ways of simulating fluid dynamical problems on a computer. Eulerian codes set up a grid of fixed boxes and compute the changes of all variables in each box as the flow evolves. Lagrangian codes instead set up an ensemble of particles which represent fluid elements and follow the trajectories of the particles in the flow. There are advantages and disadvantages to each approach which are being much discussed at present in relation to simulating the formation of stars and galaxies, two highly topical problems in astrophysical fluid dynamics. Eulerian codes in their simplest form have the disadvantage that you have to decide at the beginning of the calculation how you are going to distribute your grid cells, i.e. in what regions of the flow do you want fine gridding (high

resolution). However, as the flow evolves you might need high resolution in a region that you would not have predicted at the outset of
the calculation, unless your problem has very regular symmetry. The
Lagrangian approach circumvents this difficulty since high resolution
will automatically be achieved in high density regions of the flow
where lots of particles end up. However, the problem with using particles to model a continuous fluid is then how to compute the density
at each point (number of particles divided by some sampling volume
is an obvious approach, but then the density calculated might fluctuate
unphysically as particles enter and leave the sampling volume). Much
work has gone into devising codes that minimise the inherent noisi-
ness of particle based methods, particularly a class of codes known as
Smoothed Particle Hydrodynamics (SPH). In the meantime, Eulerian
codes are becoming more cunning through developing the capability of
reconfiguring the grid automatically during the calculation in order to
achieve high resolution where it is needed (Adaptive Mesh Refinement
methods: AMR). There is a large and evolving literature on computing
methods for fluid problems – see e.g. the Von Karman Institute Lecture
Series Monograph: *Computational Fluid Dynamics*.

1.4 Relation between the Eulerian and Lagrangian descriptions

Consider any quantity Q (which may be a scalar or a vector) in a fluid
element which is at position $\mathbf{r}$ at time t. At time $t + \delta t$ the element is
at $\mathbf{r} + \delta\mathbf{r}$, and then, straight from the definition

$$\frac{DQ}{Dt} = \lim_{\delta t \to 0} \left[\frac{Q(\mathbf{r} + \delta\mathbf{r}, t + \delta t) - Q(\mathbf{r}, t)}{\delta t} \right]. \tag{1.4}$$

The numerator is

$$Q(\mathbf{r} + \delta\mathbf{r}, t + \delta t) - Q(\mathbf{r}, t) = Q(\mathbf{r}, t + \delta t) - Q(\mathbf{r}, t)$$
$$+ Q(\mathbf{r} + \delta\mathbf{r}, t + \delta t) - Q(\mathbf{r}, t + \delta t) \tag{1.5}$$

which is, to first order in $\delta\mathbf{r}$, δt,

$$= \frac{\partial Q(\mathbf{r}, t)}{\partial t} \delta t + \delta\mathbf{r} \cdot \nabla Q(\mathbf{r}, t + \delta t), \tag{1.6}$$

and so, expanding the second term,

$$= \frac{\partial Q(\mathbf{r}, t)}{\partial t} \delta t + \delta\mathbf{r} \cdot \left[\nabla Q(\mathbf{r}, t) + \frac{\partial \nabla Q}{\partial t} \delta t \cdots \right]. \tag{1.7}$$

The $\delta\mathbf{r} \cdot \frac{\partial \nabla Q}{\partial t} \delta t$ term is of second order in small quantities, so, in the
limit as δt and $\delta r \to 0$,

$$\frac{DQ}{Dt} = \frac{\partial Q}{\partial t} + \mathbf{u} \cdot \nabla Q, \tag{1.8}$$

where $\mathbf{u}$ is the fluid velocity.

The Lagrangian time derivative has a term due to the rate of change at a fixed location (i.e. the Eulerian time derivative) plus a term due to the fact that the fluid element has moved to a new location where the variable has a different value. The extra term is known as the 'convective derivative'.

As a reminder, for scalar Q, $\mathbf{u} \cdot \nabla Q$ in Cartesian, spherical polar and cylindrical polar coordinates is

$$
\begin{aligned}
\mathbf{u} \cdot \nabla Q &= u_x \frac{\partial Q}{\partial x} + u_y \frac{\partial Q}{\partial y} + u_z \frac{\partial Q}{\partial z} \\
&= u_r \frac{\partial Q}{\partial r} + \frac{u_\theta}{r} \frac{\partial Q}{\partial \theta} + \frac{u_\phi}{r \sin\theta} \frac{\partial Q}{\partial \phi} \\
&= u_R \frac{\partial Q}{\partial R} + u_z \frac{\partial Q}{\partial z} + \frac{u_\phi}{R} \frac{\partial Q}{\partial \phi}.
\end{aligned}
$$

If $\mathbf{Q}$ is a vector, then in Equation (1.8) $\mathbf{u} \cdot \nabla \mathbf{Q}$ is also a vector, each component of which is $\mathbf{u} \cdot \nabla$ acting on each component of $\mathbf{Q}$. So, in Cartesians, where $\mathbf{Q} = (Q_x, Q_y, Q_z)$, we have

$$
\begin{aligned}
\mathbf{u} \cdot \nabla \mathbf{Q} = \Bigg(& u_x \frac{\partial Q_x}{\partial x} + u_y \frac{\partial Q_x}{\partial y} + u_z \frac{\partial Q_x}{\partial z}, \, u_x \frac{\partial Q_y}{\partial x} + u_y \frac{\partial Q_y}{\partial y} + u_z \frac{\partial Q_y}{\partial z}, \\
& u_x \frac{\partial Q_z}{\partial x} + u_y \frac{\partial Q_z}{\partial y} + u_z \frac{\partial Q_z}{\partial z} \Bigg). \tag{1.9}
\end{aligned}
$$

For these expressions in spherical polars, or cylindrical coordinates, see Appendix.

1.5 Kinematical concepts

Kinematics is the study of particle trajectories. It is distinct from dynamics in that kinematics does not concern itself with the origin of particle motions but just analyses various properties of particles moving in a known velocity field $\mathbf{u}(\mathbf{r}, t)$ (i.e. a vector field defined everywhere in Eulerian coordinates).

(i) Streamlines
The defining feature is that the tangent to a streamline at any point gives the direction of the velocity at that point. The tangent to a curve with parameter s is given in Cartesian coordinates by the vector $(\frac{dx}{ds}, \frac{dy}{ds}, \frac{dz}{ds})$, so streamlines are determined by the following system of equations:

$$\frac{dx}{u_x} = \frac{dy}{u_y} = \frac{dz}{u_z},\tag{1.10}$$

where the u are evaluated everywhere at the particular time of interest.

In spherical polars the form is slightly different:

$$\frac{dr}{u_r} = \frac{r\,d\theta}{u_\theta} = \frac{r\sin\theta\,d\phi}{u_\phi},\tag{1.11}$$

just reflecting the fact that we chose to have each component of $\mathbf{u}$ as the velocity projection in the orthogonal coordinate increment directions, $\mathbf{u} = (\frac{dr}{dt}, r\frac{d\theta}{dt}, r\sin\theta\frac{d\phi}{dt})$.

In cylindrical polars:

$$\frac{dR}{u_R} = \frac{dz}{u_z} = \frac{R\,d\phi}{u_\phi}.\tag{1.12}$$

(ii) Particle paths

The paths of individual particles as functions of time satisfy

$$\frac{d\mathbf{r}}{dt} = \mathbf{u}(\mathbf{r}, t).\tag{1.13}$$

The constant of integration labels the different particles – for example you might use $\mathbf{r}(t=0)$ to label a particle. The particle paths follow the streamlines for small times relative to the start time, since the $\mathbf{u}$ may be treated as a constant then, but if the flow is time dependent then the streamlines and the particle paths are not the same.

The equations of particle paths in various coordinate systems are as follows:

Cartesian:

$$\frac{dx}{dt} = u_x, \qquad \frac{dy}{dt} = u_y, \qquad \frac{dz}{dt} = u_z.$$

Spherical:

$$\frac{dr}{dt} = u_r, \qquad r\frac{d\theta}{dt} = u_\theta, \qquad r\sin\theta\frac{d\phi}{dt} = u_\phi.$$

Cylindrical:

$$\frac{dR}{dt} = u_R, \qquad \frac{dz}{dt} = u_z, \qquad R\frac{d\phi}{dt} = u_\phi.$$

(iii) Streaklines

A streakline is the line (at a particular time t) joining the instantaneous positions of all the particles which have ever passed (and will ever pass) through a particular point. The way to visualise this is to think of all particles passing through a given point being dyed red as they do so: the streakline is the resulting red line (for example, these are the lines you see in some wind tunnel experiments).

The equations of a streakline involve determining what subset of particles ever pass through a particular point $\mathbf{r}_0$, i.e. for what particle labels $\mathbf{a}$ is $\mathbf{r}(\mathbf{a}, t) = \mathbf{r}_0$ for some value of t? The streakline is then $\mathbf{r}(\mathbf{a}, 0)$, where $\mathbf{a}$ ranges through all the values satisfying the condition, and $\mathbf{r}_0$ labels each streakline.

Note that for *steady* flows streamlines, streaklines and particle paths are the *same*. For a steady flow, $\frac{\partial}{\partial t} = 0$, so every particle passing through a given point follows the same path.

For an example where they are different we need a non-steady flow. Suppose we have a flow where, for $t < 0$, $\mathbf{u} = (1, 0, 0)$, and for $t > 0$ $\mathbf{u} = (0, 1, 0)$. The streamlines are shown in Figure 1.2.

Fig. 1.2. Streamlines for $t < 0$ (left) and $t > 0$ (right).

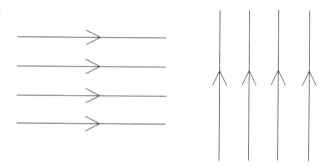

The particle paths reflect this change in velocity:

Fig. 1.3. Particle paths, where the change in direction occurs at $t = 0$.

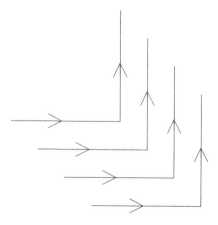

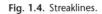

Fig. 1.4. Streaklines.

In the case of streaklines the horizontal line represents those particles which passed through some position $\mathbf{r}_p$ at $t < 0$, and the vertical
line those which passed there at $t > 0$.

Chapter 2
The fluid equations

The equations which describe the motion of fluid elements are based on concepts which are familiar from Newtonian mechanics, namely the conservation of mass, momentum and energy. So we demand that for any region the rate of change of its mass is the net flow of mass into it, that the rate of change of momentum is balanced by momentum flow and net force, and the rate of change of total energy is determined by energy gains minus losses from outside. In this chapter we use these principles to formulate the mass and momentum fluid equations. The energy equation is considered in Chapter 4.

2.1 Conservation of mass

Consider a fixed volume V whose surface S is a patchwork of surface elements dS.

If the mass density of the fluid is ρ, the rate of change of mass of the fluid contained in the volume V is

$$\frac{\partial}{\partial t} \int_V \rho \, dV. \tag{2.1}$$

In the absence of sources or sinks for matter, this must be equal to the net inflow of mass integrated over the whole surface.

The outward mass flow across an element dS is $\rho \mathbf{u} \cdot d\mathbf{S}$. To see this we suppose the velocity vector $\mathbf{u}$ at dS is at an angle θ to the surface element vector $d\mathbf{S}$ (recall that $d\mathbf{S}$ always lies along the local normal to the surface and with a magnitude equal to the area of the patch concerned). The distance travelled by the fluid particles per unit time in the direction of $d\mathbf{S}$ is then $u\cos\theta = \mathbf{u} \cdot d\mathbf{S}/|d\mathbf{S}|$. The mass of

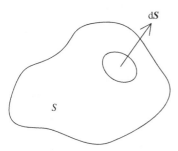

dS

S

Fig. 2.1. The surface around a fixed volume.

fluid crossing the surface in that time is the density $\rho \times$ this length $\times$ the surface area $|\mathrm{d}\mathbf{S}|$, which is $\rho\mathbf{u} \cdot \mathrm{d}\mathbf{S}$.

Hence the mass gained by the volume V is obtained by integrating over the surface (with a change of sign because the expression we have above is for an outflow):

$$-\int_S \rho\mathbf{u} \cdot \mathrm{d}\mathbf{S} = -\int_V \mathbf{\nabla} \cdot (\rho\mathbf{u})\,\mathrm{d}V, \tag{2.2}$$

where equality with the volume integral comes from the divergence theorem.

Therefore

$$\frac{\partial}{\partial t}\int_V \rho\,\mathrm{d}V = -\int_V \mathbf{\nabla} \cdot (\rho\mathbf{u})\,\mathrm{d}V. \tag{2.3}$$

Since this is true for all volumes, we can write

$$\frac{\partial\rho}{\partial t} + \mathbf{\nabla} \cdot (\rho\mathbf{u}) = 0. \qquad \textit{Eulerian} \tag{2.4}$$

This is the usual (Eulerian) form of the continuity equation.

To write this in Lagrangian form all we have to do is use the relation (1.8)

$$\frac{\mathrm{D}Q}{\mathrm{D}t} = \frac{\partial Q}{\partial t} + \mathbf{u} \cdot \mathbf{\nabla}Q, \tag{2.5}$$

which was derived earlier, along with the relation $\mathbf{\nabla} \cdot (\rho\mathbf{u}) = \rho\mathbf{\nabla} \cdot \mathbf{u} + \mathbf{u} \cdot \mathbf{\nabla}\rho$. So, in Lagrangian form, the continuity equation is

$$\frac{\mathrm{D}\rho}{\mathrm{D}t} = \frac{\partial\rho}{\partial t} + \mathbf{u} \cdot \mathbf{\nabla}\rho = -\mathbf{\nabla} \cdot (\rho\mathbf{u}) + \mathbf{u} \cdot \mathbf{\nabla}\rho = -\rho\mathbf{\nabla} \cdot \mathbf{u}, \tag{2.6}$$

i.e.

$$\frac{D\rho}{Dt} + \rho \boldsymbol{\nabla} \cdot \mathbf{u} = 0. \qquad \textit{Lagrangian} \qquad (2.7)$$

The definition of *incompressible* flows is that $D\rho/Dt = 0$. This does not imply that the density is necessarily constant everywhere, but that individual fluid elements preserve their density along their paths. For incompressible flows, therefore, $\boldsymbol{\nabla} \cdot \mathbf{u} = 0$. Incompressible flows thus have the special and useful property of being *divergence free*.

2.2 Pressure

The next thing we want to do is set up the momentum equations, which involve forces within the fluid. This means we have to consider the forces between fluid elements. In this section we will consider only *collisional* fluids where there are forces within the fluid that relate to its local temperature, through the laws of thermodynamics (see discussion in Section 1.2).

If we take any surface within a fluid, there is a momentum flux across that surface (from each side) which has nothing to do with any bulk flows in the fluid but is a consequence of its thermal properties. To understand this, it is by far the easiest to consider the behaviour of the fluid at a microscopic level. At finite temperature, molecules in a gas are in a state of random motion (i.e. with respect to the local rest frame of the fluid). The pressure is simply the (one-sided) momentum flux associated with these random motions. Obviously in a fluid with uniform properties, this momentum flux is balanced by an equal and opposite momentum flux through the other side of the hypothetical surface and there is no *net* acceleration of the fluid. The pressure is not zero in this case, however, as it is defined as the momentum flux on *one* side of the surface.

If these motions are isotropic, the momentum flux locally is independent of the orientation of the surface. Likewise, the momentum flux is always perpendicular to the surface, since the components parallel to the surface cancel out. Therefore we can write the force per unit area acting on the surface from one side as

$$\mathbf{F} = p\hat{\mathbf{s}}, \qquad (2.8)$$

where p is the pressure and $\hat{\mathbf{s}}$ is the unit normal to the surface $\mathbf{S}$.

In the more general case the forces across a surface need not be perpendicular to the surface, and so we write, in a component notation with the repeated index summation convention,

$$F_i = \sigma_{ij}\hat{s}_j, \tag{2.9}$$

where the quantities σ_{ij} are components of the *stress tensor*. (In other words, σ_{ij} is the force in the direction i acting on a surface whose normal is in direction j.) Any stress tensor can be decomposed into the sum of a diagonal tensor with equal elements and a residual tensor. The pressure is *defined* to be the elements of the diagonal tensor, which can be written $p\,\delta_{ij}$, where δ_{ij} is the Kronecker delta ($\delta_{ij} = 1$ if $i = j$ and is 0 otherwise). Thus although it is convenient to envisage pressure in terms of billiard ball bombardment as in the kinetic theory of ideal gases, *any* microscopic effects which give rise to a stress component of this form can be included in the fluid equations as a 'pressure'. (For example, in Chapter 13 we will see that some components of the forces produced by magnetic fields may be represented as an effective pressure term.)

We shall now turn to the derivation of the momentum equation involving pressure terms.

2.3 Momentum equations

Consider a lump of fluid subject to

(a) gravity with local acceleration due to gravity of $\mathbf{g}$,
(b) pressure from the surrounding fluid.

The pressure force acting on the fluid on the surface element dS of the region we are considering is $-p\,\mathrm{d}\mathbf{S}$. The minus sign arises simply because the surface element vector is outwards, and the force acting on the element is inwards.

Now the component of the force in a direction $\hat{\mathbf{n}}$ is therefore $-p\hat{\mathbf{n}} \cdot \mathrm{d}\mathbf{S}$. Hence the net force acting over the whole surface in the direction $\hat{\mathbf{n}}$ is

$$\begin{aligned}
F &= -\int_S p\hat{\mathbf{n}} \cdot \mathrm{d}\mathbf{S} \\
&= -\int_V \boldsymbol{\nabla} \cdot (p\hat{\mathbf{n}})\,\mathrm{d}V, \tag{2.10}
\end{aligned}$$

from the divergence theorem.

Fig. 2.2. A fluid volume.

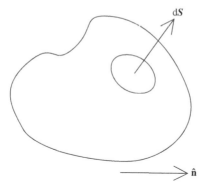

The equation of motion for the fluid element, in the direction $\hat{\mathbf{n}}$, is given by setting the rate of change of momentum for the element (using the Lagrangian derivative, because we are following the element) equal to the force in that direction. So

$$\left(\frac{D}{Dt}\int_V \rho\,\mathbf{u}\,dV\right)\cdot\hat{\mathbf{n}} = -\int_V \boldsymbol{\nabla}\cdot(p\hat{\mathbf{n}})\,dV + \int_V \rho\,\mathbf{g}\cdot\hat{\mathbf{n}}\,dV, \qquad (2.11)$$

where the second term on the right of Equation (2.11) is the gravitational force ('mg') on the element.

Note that

$$\boldsymbol{\nabla}\cdot(p\hat{\mathbf{n}}) = \hat{\mathbf{n}}\cdot\boldsymbol{\nabla}p + p\boldsymbol{\nabla}\cdot\hat{\mathbf{n}}, \qquad (2.12)$$

and the second term on the right of (2.12) is zero because $\hat{\mathbf{n}}$ is a unit vector in a constant direction. Then, in the limit when the fluid lump is small, we can replace $\int dV$ by δV, and then we have

$$\frac{D}{Dt}(\rho\,\mathbf{u}\,\delta V)\cdot\hat{\mathbf{n}} = -\delta V\,\hat{\mathbf{n}}\cdot\boldsymbol{\nabla}p + \delta V\,\rho\,\mathbf{g}\cdot\hat{\mathbf{n}}. \qquad (2.13)$$

This becomes

$$\hat{\mathbf{n}}\cdot\mathbf{u}\frac{D}{Dt}(\rho\,\delta V) + \rho\,\delta V\,\hat{\mathbf{n}}\cdot\frac{D\mathbf{u}}{Dt} = -\delta V\,\hat{\mathbf{n}}\cdot\boldsymbol{\nabla}p + \delta V\,\rho\,\mathbf{g}\cdot\hat{\mathbf{n}}, \qquad (2.14)$$

using the expansion for $\boldsymbol{\nabla}\cdot(a\,\mathbf{b})$ again. The first differential is the rate of change of mass of the lump of fluid we are following. Since mass is conserved, this is zero, so we are left with

$$\rho\,\delta V\,\hat{\mathbf{n}}\cdot\frac{D\mathbf{u}}{Dt} = -\delta V\,\hat{\mathbf{n}}\cdot\boldsymbol{\nabla}p + \delta V\,\rho\,\mathbf{g}\cdot\hat{\mathbf{n}}. \qquad (2.15)$$

This is true for all δV and $\hat{\mathbf{n}}$, and so finally we have

$$\rho \frac{D\mathbf{u}}{Dt} = -\nabla p + \rho \mathbf{g}. \qquad \textit{Lagrangian} \qquad (2.16)$$

This is the Lagrangian form of the momentum equation: the momentum of a fluid element changes as a result of pressure *gradients* and gravitational forces.

To transform to the Eulerian form we use the relation (1.8), and then

$$\rho \frac{\partial \mathbf{u}}{\partial t} + \rho \mathbf{u} \cdot \nabla \mathbf{u} = -\nabla p + \rho \mathbf{g}. \qquad \textit{Eulerian} \qquad (2.17)$$

This says that the momentum contained in a fixed grid cell within the fluid changes as a result of pressure and gravitational forces, and any imbalance in the momentum flux in and out of the grid cell.

2.4 Momentum equation in conservative form: the stress tensor and concept of ram pressure

We may get a different insight into the meaning of this equation by instead evaluating the (Eulerian) rate of change of momentum density $(\rho \mathbf{u})$. Here we use component notation with the summation convention (sum over repeated indices), in Cartesian coordinates:

$$\frac{\partial}{\partial t}(\rho u_i) = \rho \frac{\partial u_i}{\partial t} + u_i \frac{\partial \rho}{\partial t}$$
$$= -\rho u_j \partial_j u_i - \partial_j p \delta_{ij} + \rho g_i - u_i \partial_j (\rho u_j), \qquad (2.18)$$

where $\partial_j \equiv \frac{\partial}{\partial x_j}$, where x_j is the jth coordinate variable, and δ_{ij} is the Kronecker delta. The first three terms come from the momentum equation, and the last from that of mass conservation. Introducing the δ_{ij} may seem unnecessary here, but it allows us to write the equations as

$$\frac{\partial}{\partial t}(\rho u_i) = -\partial_j(\rho u_j u_i + p \delta_{ij}) + \rho g_i. \qquad (2.19)$$

In order to see the physical content of Equation (2.19), consider the flow:

The mass per second hitting the surface with $\hat{\mathbf{s}}$ in the j direction due to the bulk flow is ρu_j. Thus the momentum flux per second in the i direction through a surface with normal in the j direction is $\rho u_j u_i$.

Fig. 2.3. Flow components.

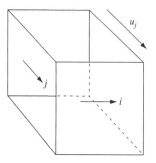

(By symmetry, this is also equal to the momentum flux per second in the j direction through a surface with normal in the i direction.) Recalling the definition of the stress tensor σ_{ij}, i.e. the force in direction i on unit surface with normal in direction j,

$$F_i = \sigma_{ij}\,\hat{s}_j, \tag{2.20}$$

we see that $\rho u_j u_i$ gives the components of the stress tensor σ_{ij} due to bulk flows. The total stress tensor is then the sum of this and the diagonal tensor $p\delta_{ij}$ provided by the thermal pressure.

Therefore we can write the momentum equation (in conservative form) as

$$\frac{\partial}{\partial t}(\rho u_i) = -\partial_j\sigma_{ij} + \rho g_i, \tag{2.21}$$

where $\sigma_{ij} = \rho u_j u_i + p\delta_{ij}$, and σ_{ij} is defined by the force equation $F_i = \sigma_{ij}\hat{s}_j$.

Going back to Equation (2.19) we see that the term $\partial_j(\rho u_j u_i)$ represents the change in the ith component of momentum due to a mismatch in 'i-momentum' carried through a cell in each of the three orthogonal directions. For example, in Cartesian coordinates

$$\partial_j(\rho u_j u_x) = \frac{\partial}{\partial x}(\rho(u_x)^2) + \frac{\partial}{\partial y}(\rho u_y u_x) + \frac{\partial}{\partial z}(\rho u_z u_x). \tag{2.22}$$

Note that $\rho u_i u_j$ has the dimensions of pressure, and is usually termed the 'ram pressure'.

The basic difference between the ram pressure and the thermal pressure is that the ram pressure is associated with bulk motions **u** whereas the thermal pressure is associated with random motions $\tilde{\mathbf{u}}$ which are *isotropic*. In the case of isotropic random motions the local average of $\rho\tilde{u}_i\tilde{u}_j$ is $\rho(\tilde{u}_j)^2$ for all j because the average $\overline{\tilde{u}_i\tilde{u}_j} = 0$ for

$j \neq i$. Therefore the thermal pressure is independent of i and j and is just a scalar. Physically, this means that thermal pressure (unlike ram pressure) always acts perpendicular to any surface in the fluid (see Section 2.2). In general $\rho u_i u_j$ does depend on i and j for bulk motions, so the ram pressure is a 'stress tensor' – a 3×3 matrix of elements $\rho u_i u_j$.

We conclude with a simple example of the flow of a hot fluid, pressure p, down a pipe. *Any* surface in the fluid will experience momentum flux p due to the thermal pressure, but only a surface whose normal has some component along the direction of the flow experiences the ram pressure.

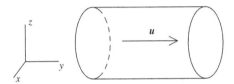

Fig. 2.4. Flow in a pipe.

The stress tensor is

$$\begin{pmatrix} p & 0 & 0 \\ 0 & p+\rho u^2 & 0 \\ 0 & 0 & p \end{pmatrix}$$

so the pressure on the sides of the pipe is p, and the pressure on the end $p+\rho u^2$.

Chapter 3
Gravitation

The gravitational force is of course of central importance in astrophysics. Therefore whereas the force of gravity plays only a minor role in most textbooks on terrestrial fluid mechanics, with such fluids being only subject to the uniform gravitational acceleration of the Earth, the astronomical situation is quite different. We will need routinely to be able to solve for the gravitational field produced by the fluid itself before we can calculate the gravitational term in the momentum equation which was derived in Chapter 2. This chapter therefore represents an aside on how to calculate gravitational fields from given density distributions.

3.1 The gravitational potential

Any force that is *conservative* (i.e. one in which the work done in a closed loop $\oint \mathbf{F} \cdot d\boldsymbol{\ell} = 0$) can be written as

$$\mathbf{F} = \boldsymbol{\nabla} \Phi, \tag{3.1}$$

where Φ is a scalar potential. We can see that this is the case by using Stokes' theorem, i.e. $\oint \mathbf{F} \cdot d\boldsymbol{\ell} = \int_S \boldsymbol{\nabla} \wedge \mathbf{F} \cdot d\mathbf{S}$. Since $\oint \mathbf{F} \cdot d\boldsymbol{\ell} = 0$ for conservative force fields then $\boldsymbol{\nabla} \wedge \mathbf{F}$ must be zero. Since 'curl of grad' is always zero, any force field of the form given in (3.1) must be conservative. This is simply a mathematical formulation of the well-known result that every point in a conservative force field can be labelled by a scalar potential function, whose gradient gives the magnitude of the force at each point.

In the case of gravity we define a (scalar) gravitational potential Ψ such that the gravitational acceleration $\mathbf{g}$ is given by

$$\mathbf{g} = -\boldsymbol{\nabla} \Psi. \tag{3.2}$$

(The minus sign in the above equation ensures that the gravitational force acts in the direction of diminishing potential, i.e. valleys have lower potential than hills.)

The work required to take a unit mass to infinity is

$$-\int_{\mathbf{r}}^{\infty} \mathbf{g} \cdot d\boldsymbol{\ell} = \int_{\mathbf{r}}^{\infty} \nabla \Psi \cdot d\boldsymbol{\ell} = \Psi(\infty) - \Psi(\mathbf{r}), \qquad (3.3)$$

since the result is independent of the path taken. $\Psi(\infty)$ is the value of the potential so far from any gravitating mass that the gravitational acceleration is zero there. It is single valued for all applications we are interested in, and is often taken as the reference zero for the potential function Ψ. Any convenient reference point will do (and infinity is as good as any), since it is only potential differences and gradients that have physical significance.

Example

Consider $\Psi = -\frac{GM}{r}$, so the potential is a function only of radius from the origin. Then

$$\nabla \Psi = -GM \nabla \left(\frac{1}{r}\right) = \frac{GM}{r^2} \left(\frac{\partial r}{\partial x}, \frac{\partial r}{\partial y}, \frac{\partial r}{\partial z}\right). \qquad (3.4)$$

Now $r^2 = x^2 + y^2 + z^2$, so $2r\frac{\partial r}{\partial x} = 2x$ and similar for y and z. Then

$$\nabla \Psi = \frac{GM}{r^3}(x, y, z) \equiv \frac{GM}{r^3}\mathbf{r} \equiv \frac{GM}{r^2}\hat{\mathbf{r}}, \qquad (3.5)$$

where $\hat{\mathbf{r}}$ is the unit radial vector at the point $(x, y, z) \equiv (r, \theta, \phi)$. Hence

$$\mathbf{g} = -\frac{GM}{r^3}\mathbf{r} = -\frac{GM}{r^2}\hat{\mathbf{r}}, \qquad (3.6)$$

i.e. a vector of modulus $\frac{GM}{r^2}$ directed towards the origin. So $\Psi = -GM/r$ is the potential of a point mass located at the origin.

Likewise, for a point mass located at $\mathbf{r}'$, the potential is

$$\Psi = -\frac{GM}{|\mathbf{r} - \mathbf{r}'|}, \qquad (3.7)$$

and, for a system of point masses M_i at locations $\mathbf{r}'_i$, the potential at $\mathbf{r}$ is

$$\Psi = -\sum_i \frac{GM_i}{|\mathbf{r} - \mathbf{r}'_i|}. \qquad (3.8)$$

The straight sum is appropriate because the forces (or accelerations) from the ∇ of the individual contributions add vectorially and because ∇ is a linear operator.

If we are given a potential then we can work out the acceleration due to gravity, but we really want to be able to determine the potential from a given mass distribution. We can obviously do this by summing, or integrating, over mass elements, and now use this to derive a general differential equation linking the potential Ψ and the density ρ.

3.2 Poisson's equation

Consider a point P surrounded by a surface S. The definition of the solid angle subtended at P by $d\mathbf{S}$ is

$$d\Omega = \frac{d\mathbf{S} \cdot \hat{\mathbf{r}}}{r^2}. \tag{3.9}$$

If we integrate over the surface

$$\int \frac{\hat{\mathbf{r}} \cdot d\mathbf{S}}{r^2} = \begin{cases} 4\pi & \text{if P is anywhere inside } S, \\ 0 & \text{if P is anywhere outside } S. \end{cases} \tag{3.10}$$

We may derive Equation (3.10) by placing the origin of coordinates at P. The left hand side may be written $\int \mathbf{f} \cdot d\mathbf{S}$ where, in Cartesian coordinates, $\mathbf{f} = (x\hat{\mathbf{x}} + y\hat{\mathbf{y}} + z\hat{\mathbf{z}})/(x^2 + y^2 + z^2)^{3/2}$. Provided that $\mathbf{f}$ is finite over the volume enclosed by S (i.e. if P is outside S), we can use the divergence theorem $\int \mathbf{f} \cdot d\mathbf{S} = \int \nabla \cdot \mathbf{f} \, dV$. It is then an easy matter to evaluate $\nabla \cdot \mathbf{f}$ in Cartesian coordinates and demonstrate that it is equal to zero, thus proving the second identity. In the case that P is inside S, then $\mathbf{f}$ is undefined at P, so in order to apply the divergence theorem we need to place a small interior surface, S', around P so that P is excluded from the volume of integration. We may now write

$$\int_S \frac{\hat{\mathbf{r}} \cdot d\mathbf{S}}{r^2} + \int_{S'} \frac{\hat{\mathbf{r}} \cdot d\mathbf{S}}{r^2} = 0. \tag{3.11}$$

Fig. 3.1. Point P inside a surface S.

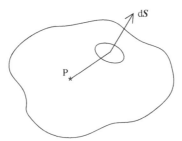

Now we are free to choose S' as we please so long as it excludes P, and Equation (3.11) is still satisfied. So, for example, we could choose S' to be a sphere of radius a. In this case, $\hat{\mathbf{r}} \cdot d\mathbf{S}/r^2 = -4\pi a^2/a^2 = -4\pi$ (the negative sign results from the fact that the outward normal from S' is in the inward radial direction with respect to P). Hence we obtain the result $\int_S (\hat{\mathbf{r}} \cdot d\mathbf{S}/r^2) = 4\pi$ if P is inside S.

Now suppose there is a mass M at the point P inside S. Since $\mathbf{g} = -(GM/r^2)\hat{\mathbf{r}}$, $\mathbf{g} \cdot d\mathbf{S} = -GMd\Omega$, and since GM is a constant we must have

$$\int_S \mathbf{g} \cdot d\mathbf{S} = -4\pi GM, \qquad (3.12)$$

if M is inside S.

If we distribute masses throughout the volume enclosed by S, then

$$\int_S \mathbf{g} \cdot d\mathbf{S} = -4\pi G \sum_i M_i = -4\pi G \int_V \rho \, dV. \qquad (3.13)$$

From the divergence theorem we therefore have

$$\int_S \mathbf{g} \cdot d\mathbf{S} = \int_V \nabla \cdot \mathbf{g} \, dV, \qquad (3.14)$$

and so

$$\int_V (\nabla \cdot \mathbf{g} + 4\pi G\rho) \, dV = 0. \qquad (3.15)$$

Since this is true for any volume, we must have

$$\nabla \cdot \mathbf{g} + 4\pi G\rho = 0, \qquad (3.16)$$

i.e.

$$\nabla \cdot (-\nabla \Psi) + 4\pi G\rho = 0, \qquad (3.17)$$

or

$$\nabla^2 \Psi = 4\pi G\rho. \qquad (3.18)$$

This is Poisson's equation. It relates any mass density distribution to the resultant potential. Since $\mathbf{g} = -\nabla \Psi$ we then have the acceleration due to gravity everywhere, which can be substituted into the right hand side of the momentum equation. Note that since $\nabla^2 \Psi \propto \rho$ the potential is a minimum in mass concentrations, as we expect from the definition of the potential function in (3.1).

3.3 Using Poisson's equation

With many problems, the most useful thing to do is to use whatever symmetries there are to simplify it. In some cases it is more appropriate to use the form (3.12), choosing the closed two-dimensional surface over which the integral is performed (often called a 'Gaussian surface') to make use of the symmetries. Choosing a Gaussian surface is not difficult, but can be subtle. There are two components, the magnitude of $\mathbf{g}$ in Equation (3.12) and its direction. To simplify the integral we try to choose a surface over which the magnitude of $\mathbf{g}$ is constant and where the normal to the surface is parallel to $\mathbf{g}$ where possible. If that proves not to be possible, then a surface which satisfies these criteria over one part, and for which $\mathbf{g}$ is parallel to a line within the surface so that $\mathbf{g} \cdot \mathbf{dS} = 0$ for the remainder, will also allow the integral to be evaluated simply.

3.3.1 A spherically symmetric mass distribution

Choose the Gaussian surface as a sphere centred on the mass distribution, and then, by symmetry

(a) $\mathbf{g}$ is radial, and
(b) the magnitude of $\mathbf{g}$ is the same everywhere on the sphere.

Then

$$\int \mathbf{g} \cdot \mathbf{dS} = -4\pi r^2 |\mathbf{g}|, \tag{3.19}$$

so

$$|\mathbf{g}| = \frac{GM(r)}{r^2} = \frac{G \int_0^r 4\pi\rho(r')\,(r')^2\,dr'}{r^2}, \tag{3.20}$$

where $M(r)$ is the enclosed mass and

$$\mathbf{g} = -|\mathbf{g}|\hat{\mathbf{r}}. \tag{3.21}$$

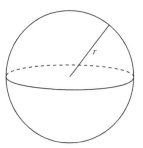

Fig. 3.2. Spherically symmetric mass distribution.

> For a spherical mass distribution **g** depends only on the
> enclosed mass.

3.3.2 An infinite cylindrically symmetric mass distribution

Use cylindrical polars with $R = 0$ as the line of symmetry (since it is a cylindrical problem!), and here **g** is uniform and radial on the curved sides of the cylindrical Gaussian surface, and zero on the flat ends (by symmetry).

Therefore, with R as the radius of the cylindrical surface, ℓ its length, and M the enclosed mass, we have

$$\int \mathbf{g} \cdot \mathrm{d}\mathbf{S} = -2\pi R\ell|\mathbf{g}| = -4\pi GM = -4\pi G \int_0^R 2\pi R' \rho(R')\ell \, \mathrm{d}R', \quad (3.22)$$

$$\Rightarrow |\mathbf{g}| = \frac{2G}{R} \int_0^R 2\pi R' \rho(R') \mathrm{d}R', \quad (3.23)$$

and

$$\mathbf{g} = -|\mathbf{g}|\hat{\mathbf{R}}. \quad (3.24)$$

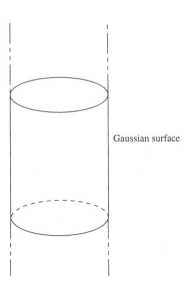

Fig. 3.3. Infinite cylindrically symmetric mass distribution.

Gaussian surface

3.3.3 An infinite planar distribution: $\rho = \rho(z)$ symmetric about $z = 0$

Fig. 3.4. A Gaussian surface
for an infinite plane mass
distribution symmetric about
$z = 0$

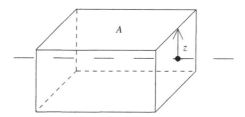

Choose the Gaussian surface as a box, area A and height $2z$. **g** is zero
on the sides of the box, so what is left is

$$\int \mathbf{g} \cdot d\mathbf{S} = -2|\mathbf{g}|A = -4\pi GM = -4\pi GA \int_{-z}^{z} \rho(z')dz'. \tag{3.25}$$

So

$$|\mathbf{g}| = 4\pi G \int_{0}^{z} \rho(z')dz' \tag{3.26}$$

(4π because the integral limit changed from $-z$ to 0), and

$$\mathbf{g} = -|\mathbf{g}|\hat{\mathbf{z}}. \tag{3.27}$$

Note that for a plane distribution of finite height z_{max}, **g** is constant
for all $z \geq z_{max}$.

3.3.4 A finite axisymmetric disc, symmetric about $z = 0$

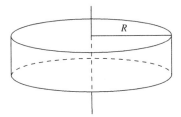

Fig. 3.5. Finite disc.

One might try in this case to take the Gaussian surface as a flat cylinder, since that is the geometry of the gravitating object. Then

$$\int \mathbf{g} \cdot d\mathbf{S} = ???$$ (3.28)

The problem has a high degree of symmetry, but there are no surfaces where $\mathbf{g}$ vanishes by symmetry arguments, and $|\mathbf{g}|$ is not uniform over the top and bottom surfaces. The relative contributions to $\int \mathbf{g} \cdot d\mathbf{S}$ from top/bottom and sides are not obvious.

This is an example where the Gaussian surface method is not useful. Although $\int \mathbf{g} \cdot d\mathbf{S} = -4\pi GM$ is always true, it is not true for the surface we have chosen that $\mathbf{g}$ locally depends only on the enclosed mass. There will be surfaces over which $|\mathbf{g}|$ is constant (not necessarily the equipotential surfaces, though, since $|\mathbf{g}| = |\nabla\Psi|$), but it is less than obvious what they are. Obtaining $\mathbf{g}$ for a disc density distribution involves the use of Bessel functions and is beyond the scope of this book.

Note that this problem occurs when the potential arises from a self-gravitating matter distribution that is disc-like. There is no such problem for discs where almost all of the potential comes from a central point mass (such as for the primitive solar nebula).

For a discussion of gravitational potentials for a range of configurations see Chapter 2 in Binney and Tremaine, *Galactic Dynamics* (Princeton University Press, 1994).

3.4 The potential associated with a spherical mass distribution

Before leaving gravitational potentials in their own right, we will look at the potential associated with a spherical mass distribution in more detail. We have only a radial component of the acceleration g, where

$$g = -\frac{G\int_0^r 4\pi\rho(r')r'^2\,\mathrm{d}r'}{r^2} = -\frac{\partial\Psi}{\partial r} = -\frac{\mathrm{d}\Psi}{\mathrm{d}r}, \quad (3.29)$$

$$\Rightarrow \Psi = \int_\infty^r \frac{G}{r''^2}\left\{\int_0^{r''} 4\pi\rho(r')r'^2\,\mathrm{d}r'\right\}\mathrm{d}r'', \quad (3.30)$$

since we have, by convention, set $\Psi(\infty) = 0$. Integrating by parts gives

$$\Psi = \left[-\frac{G}{r''}\left\{\int_0^{r''} 4\pi\rho(r')r'^2\,\mathrm{d}r'\right\}\right]_\infty^r + \int_\infty^r 4\pi G\rho(r'')r''\,\mathrm{d}r'', \quad (3.31)$$

provided $M(r)/r \to 0$ as $r \to \infty$.

Therefore, even in the spherical case, $\Psi(r)$ is affected by matter outside r: the presence of more material outside r lowers the potential at r, because we have decided to take $r = \infty$ as the reference point and it then requires more energy to take unit mass from r to ∞. So, under these circumstances, $\Psi \neq -G\,M(r)/r$ unless there is no material outside r. The outside material does not affect the potential difference between r and some point inside r, since it merely provides a zero shift in the potential for all points at distances $\leq r$ from the centre. The fact that the presence of exterior mass in a spherically symmetric distribution does not affect the acceleration but *does* affect the potential is a common point of confusion for students.

3.5 Gravitational potential energy

We have already seen (3.8) that for a system of point masses, the gravitational potential is given by

$$\Psi = -\sum_i \frac{GM_i}{|\mathbf{r} - \mathbf{r}_i|}, \quad (3.32)$$

and that $-\psi$ is the energy required to take unit mass to infinity. Thus the energy to take a system of point masses to infinity ($= -\Omega$ where Ω is the gravitational potential energy) is

$$\Omega = -\frac{1}{2}\sum_{j\neq i}\sum_i \frac{GM_iM_j}{|\mathbf{r}_j - \mathbf{r}_i|} = \frac{1}{2}\sum_j M_j\Psi_j. \quad (3.33)$$

We have introduced the factor of 1/2 here in order to ensure that each pair of masses only contributes once to the energy sum. By analogy, we may write (for a continuous matter distribution)

$$\Omega = \frac{1}{2} \int \rho(\mathbf{r})\Psi(\mathbf{r}) \, dV \qquad (3.34)$$

We can confirm that this is the correct expression in the case of a spherical mass distribution as follows. In this case, Equation (3.34) implies

$$\Omega = \frac{1}{2} \int_0^\infty 4\pi\rho(r')r'^2\Psi(r') \, dr'. \qquad (3.35)$$

Integrating by parts gives

$$\frac{1}{2}\left([M(r')\Psi(r')]_0^\infty - \int_0^\infty M(r')\frac{d\Psi}{dr'}dr' \right). \qquad (3.36)$$

The first term is zero since $M(0) = 0$ (and $\Psi(0)$ is finite) and $\Psi(\infty) = 0$ (and $M(\infty)$ is finite). For the second term we use $d\Psi/dr = GM(r)/r^2$, and so the total potential energy of the mass distribution becomes

$$-\frac{1}{2} \int_0^\infty \frac{GM(r')^2}{r'^2}dr'. \qquad (3.37)$$

Integrating by parts gives

$$\frac{1}{2}\left[G\frac{M(r')^2}{r'} \right]_0^\infty - G\int_0^\infty \frac{M(r')}{r'}\frac{dM}{dr'}dr'. \qquad (3.38)$$

The first term is zero for similar reasons as before, and we are left with

$$\Omega = -G \int_0^\infty \frac{M(r')}{r'}dM(r'). \qquad (3.39)$$

We can see that this is indeed the expected expression for the gravitational potential energy by considering the assembly of a spherical mass distribution through the bringing in of successive shells from infinity. As each shell is brought in, it loses a potential energy $(GM(r')/r') \, dM(r')$, so the total energy released in this process is $\int (GM(r')/r') \, dM(r')$.

3.6 The virial theorem

Consider now the general motion of a cloud of particles, where the particles can be anything from atoms to galaxies, depending on the context.

If a particle has a mass m, and is at $\mathbf{r}$, and is acted upon by a force $\mathbf{F}$, then we know

$$m\frac{d^2\mathbf{r}}{dt^2} = \mathbf{F}. \tag{3.40}$$

We can proceed by examining the second derivative of the moment of inertia of the particle about the origin.

$$
\begin{aligned}
\frac{1}{2}\frac{d^2(mr^2)}{dt^2} &= m\frac{d}{dt}\left(\mathbf{r}\cdot\frac{d\mathbf{r}}{dt}\right) \\
&= m\mathbf{r}\cdot\frac{d^2\mathbf{r}}{dt^2} + m\left(\frac{d\mathbf{r}}{dt}\right)^2 \\
&= m\left(\frac{d\mathbf{r}}{dt}\right)^2 + \mathbf{r}\cdot\mathbf{F}.
\end{aligned}
\tag{3.41}
$$

The first term on the right hand side is simply twice the kinetic energy of the particle. Hence, summing this over all particles we have

$$\frac{1}{2}\frac{d^2 I}{dt^2} = 2T + \sum(\mathbf{r}\cdot\mathbf{F}), \tag{3.42}$$

where I is the moment of inertia about the origin defined by $I = \sum(mr^2)$, and T is the total kinetic energy of motion of the particles forming the cloud. The quantity $V = \sum(\mathbf{r}\cdot\mathbf{F})$ is called the 'virial', since R. Clausius called it that when he first obtained this relationship in 1870.

For this to be useful, we need to evaluate V. To do this consider any two particles of masses m_i and m_j, at points $\mathbf{r}_i$ and $\mathbf{r}_j$. If the force on the second exerted by the first is $\mathbf{F}_{j(i)}$, then the force on the first exerted by the second is $\mathbf{F}_{i(j)} = -\mathbf{F}_{j(i)}$. The contribution of this pair of forces to the virial is then

$$\mathbf{F}_{i(j)}\cdot(\mathbf{r}_i - \mathbf{r}_j), \tag{3.43}$$

and hence, in the absence of an external force field,

$$V = \sum_i\sum_{j>i}\mathbf{F}_{i(j)}\cdot(\mathbf{r}_i - \mathbf{r}_j), \tag{3.44}$$

where the summation is over all pairs of particles (which we have numbered in some arbitrary order, and set $j > i$ to avoid double-counting). If the ideal gas laws apply (so collisional processes, if they occur at all, occur when $\mathbf{r}_i - \mathbf{r}_j = 0$; see Section 4.1) then all forces except gravitational forces may be neglected. Then the force $\mathbf{F}_{i(j)}$ is

$$-G\frac{m_i m_j}{r_{ij}^3}\mathbf{r}_{ij}, \tag{3.45}$$

where $\mathbf{r}_{ij} \equiv \mathbf{r}_i - \mathbf{r}_j$, and r_{ij} is the length of that vector. Therefore

$$V = -\sum_i \sum_{j>i} \frac{G m_i m_j}{r_{ij}}. \tag{3.46}$$

Now each term is simply the work done in separating the pair of particles to infinity against the gravitational attraction. Thus V is seen to be the potential energy, Ω, of the cloud of particles under consideration. Hence we have

$$\frac{1}{2}\frac{d^2 I}{dt^2} = 2T + \Omega. \tag{3.47}$$

If the system is in a steady state, I is constant, and consequently we have

$$2T + \Omega = 0, \tag{3.48}$$

which expresses the virial theorem in its most commonly used form.

In general, T is the total kinetic energy of all the particles (atoms or stars . . .) in the system. In the case of *fluids*, it is sometimes convenient to decompose this total kinetic energy into the kinetic energy of the mean flow locally and the additional kinetic energy stored in particle motions in the rest frame of the local fluid element. This latter is the *thermal* energy of the fluid in the case of a monatomic gas, which we denote by U (in chapter 4 we use $\mathcal{E}$ for the thermal, or internal, energy per unit mass). We may therefore re-write the virial theorem (in the case that the ratio of specific heats is equal to γ) in the form:

$$2T_k + 3(\gamma - 1)U + \Omega = 0 \tag{3.49}$$

where T_k is now the total kinetic energy contained in the mean streaming motion of the particles at every point.

Chapter 4
The energy equation

So far, we have derived two Eulerian equations which describe the motion of a fluid,

$$\frac{\partial \rho}{\partial t} + \boldsymbol{\nabla} \cdot (\rho \mathbf{u}) = 0, \tag{4.1}$$

$$\rho \frac{\partial \mathbf{u}}{\partial t} + \rho \mathbf{u} \cdot \boldsymbol{\nabla} \mathbf{u} = -\boldsymbol{\nabla} p - \rho \boldsymbol{\nabla} \Psi. \tag{4.2}$$

To close these we need relations which give us Ψ and p in terms of the other variables ρ and $\mathbf{u}$. We know in principle how to get Ψ, since we have $\nabla^2 \Psi = 4\pi G \rho$ (see Chapter 3), but we still have to find some relationship which allows us to determine p. The relationship between p and other thermodynamic properties of the system is called the *equation of state*. As emphasised in Chapter 1, one can only talk about an equation of state in the case of a *collisional* fluid (e.g. a gas of interacting particles) since thermodynamic relations presuppose that the components can interact in order to achieve the most likely (maximum entropy) state.

4.1 Ideal gases

Most of the fluids in the Universe can be approximated as *ideal* gas. At a microscopic level, ideal gases are those that are well described in terms of kinetic theory, in which the gas is modelled as an ensemble of point-like particles with isotropic random motions at an rms level set by the gas temperature. Although the isotropy of particle motions and the thermodynamically controlled distribution of particle speeds imply that the particles must interact at some level, these interactions are very infrequent in an ideal gas. The fact that long range forces between particles may be neglected for most purposes means that the internal energy of ideal gases is almost entirely contained in kinetic energy and is thus a function only of the temperature.

It is known from laboratory experiments that gases only show approximately ideal behaviour when they are relatively warm and dilute. If one cools or compresses an ideal gas sufficiently, the potential energy of interparticle interactions becomes important compared with the kinetic energy of random motion. This modifies the equation of state and means that the internal energy is a function of density as well as temperature (if interparticle repulsive forces are important, for example, then the internal energy is raised at high density). Likewise, the finite size of atoms and molecules becomes important at high density.

As we stressed in Chapter 1, although the gas in the Universe is very diverse in its properties, it is almost all extremely dilute by terrestrial standards and therefore the ideal gas condition is readily met. An exception to this is in the interior of giant planets where the high pressures and densities imply a significant deviation from ideal behaviour (and where, in fact, the equation of state is not well understood). Another place where the ideal gas equation of state is not appropriate is in the interiors of neutron stars and white dwarfs where the densities are such that the distribution of particle energies becomes restricted by quantum mechanical requirements on the number of particles that can populate a given energy level. We will need to treat the equation of state in this case when modelling the internal structure of neutron stars and white dwarfs. For the remainder of this chapter, however, we will mainly be discussing ideal gases.

The equation of state of an ideal gas is given by $p = \frac{\mathcal{R}_*}{\mu}\rho T$, where $\mathcal{R}_*$ is the modified gas constant and μ the mean molecular weight of the constituents of the gas. Throughout this book we use the modified gas constant which is simply $1000 \times$ the usual gas constant ($\mathcal{R} = 8.3$ joules per kelvin per mole, i.e. 8.3 joules per kelvin per molecular weight of the substance in grams). Instead we have $\mathcal{R}_* = 8300$ joules per molecular weight of the substance in *kilograms*. With this definition, we need not worry about moles any more, but can work straightforwardly with masses of fluids expressed in the SI unit of kilograms.

4.2 Barotropic equations of state: the isothermal and adiabatic cases

It may seem that we are no closer to finding the pressure of gas at a given density since we have just, through our ideal gas equation of state, expressed the pressure in terms of a further unknown, the temperature.

The temperature relates to the *internal energy content of the fluid* which is in turn determined by an equation that takes account of the various heat input and heat loss mechanisms in the fluid. We shall derive this energy equation in Section 4.3. However, under certain circumstances it may not be necessary to solve the energy equation, since one is in a physical limit where the dependence of the temperature on density may be clear. Under these special circumstances, one by-passes the energy equation and instead just uses the equation of state in this special form in order to prescribe the dependence of pressure on density in the momentum equation. Equations of state in which pressure is a function of density *only* are known as *barotropic* equations of state. We will consider two well-known examples.

An isothermal equation of state implies that temperature T is a constant, so (for an ideal gas) $p \propto \rho$. For the isothermal approximation to be a good one it is necessary that in *thermal equilibrium* (i.e. when heating balances cooling), the heating and cooling processes thermostatically control the temperature to lie within a narrow range. For time-dependent problems, it is also necessary that the system can relax to this constant temperature thermal equilibrium on timescales that are short compared with the flow times.

The other limit in which one can readily prescribe a barotropic equation of state is where an ideal gas undergoes *reversible* changes (see below) which are *adiabatic*, i.e. thermally isolated from their surroundings. In order to derive the equation of state in this case, however, we first need to develop some elementary thermodynamic relations.

We shall start with the the first law of thermodynamics, which is an expression of energy conservation:

$$\mathrm{d}Q = \mathrm{d}\mathcal{E} + p\,\mathrm{d}V. \tag{4.3}$$

Here $\mathrm{d}Q$ is the quantity of heat absorbed by unit mass of fluid from its surroundings, $p\,\mathrm{d}V$ is the work done by unit mass of fluid if its volume changes by $\mathrm{d}V$ and $\mathrm{d}\mathcal{E}$ is the change in the *internal energy* content of unit mass of the fluid. We note that this law is only valid if one can neglect processes (termed viscous or dissipative processes) that can convert the kinetic energy of the fluid into heat. In the more general case, where viscosity cannot be neglected, we have $\mathrm{d}Q < \mathrm{d}\mathcal{E} + p\,\mathrm{d}V$ because extra heat can be fed into the fluid through dissipation of its kinetic energy. In this book, we encounter dissipative processes in the case of both shocks (Chapter 7) and accretion discs (Chapter 12). For now, however, we neglect viscous dissipation (i.e. we only consider what are termed *reversible* changes) and proceed with the first law in the form (4.3).

Conventionally the changes in heat absorbed and work done are expressed using Pfaffian operators (đ), which denote that the change in these quantities in going from an initial state to a final one depends on the route taken through thermodynamic phase space. The change in internal energy is denoted by an ordinary differential because the internal energy is an inherent property of the fluid which depends only on the current values of its thermodynamic variables.

For an ideal gas (i.e. one whose equation of state is given by $p = \frac{\mathcal{R}_*}{\mu}\rho T$), it can be shown that $\mathcal{E} = \mathcal{E}(T)$, so the energy equation can be written

$$\text{đ}Q = \frac{\text{d}\mathcal{E}}{\text{d}T}\text{d}T + p\,\text{d}V \tag{4.4}$$

or, alternatively, in the form

$$\text{đ}Q = C_V\,\text{d}T + \frac{\mathcal{R}_*T}{\mu V}\,\text{d}V, \tag{4.5}$$

where the specific heat capacity at constant volume is defined to be

$$C_V = \frac{\text{d}\mathcal{E}}{\text{d}T}. \tag{4.6}$$

Thus in the case of a gas undergoing a reversible adiabatic change we have

$$C_V\,\text{d}T + \frac{\mathcal{R}_*T}{\mu V}\,\text{d}V = 0, \tag{4.7}$$

i.e.

$$C_V\,\text{d}\ln T + \frac{\mathcal{R}_*}{\mu}\,\text{d}\ln V = 0. \tag{4.8}$$

This implies that

$$V \propto T^{-\frac{C_V}{\mathcal{R}_*/\mu}} \tag{4.9}$$

and thus, substituting for the ideal gas equation of state, we also obtain the scalings:

$$p \propto T^{1+\frac{C_V}{\mathcal{R}_*/\mu}} \tag{4.10}$$

and

$$p \propto V^{-1+\frac{\mathcal{R}_*}{\mu C_V}}. \tag{4.11}$$

The actual value of C_V depends on the number of ways that the gas can store kinetic energy. We will be using without proof the well-known thermodynamic result that the specific heat capacity (per kg) of an ideal gas is $\mathcal{R}_*/2\mu$ times the number of degrees of freedom in the gas, i.e. the number of independent energy terms involving quadratic functions of phase space coordinates such as position or velocity. Evidently, if we are dealing with ideal gases, then the energy associated with the centre of mass motion of each molecule is entirely kinetic, and this motion contributes three degrees of freedom (one for the kinetic energy in each orthogonal direction for molecules moving in three dimensions). If the gas is monatomic, then this is the only way that the gas can store energy and hence the specific heat capacity at constant volume is simply $3\mathcal{R}_*/2\mu$ and the corresponding expression for the internal energy (per kg) is $3\mathcal{R}_* T/2\mu$. For more complex molecules there are additional ways that the gas can store energy: in kinetic energy of internal rotation and vibration and in potential energy associated with interatomic attractive and repulsive forces within the molecule. At any temperature, each of the independent kinetic and potential energy terms associated with molecular vibration and rotation contributes a further $\mathcal{R}_*/2\mu$ to the specific heat capacity at constant volume, *provided that these motions are excited at the temperature concerned.* This last statement is a reference to the fact that molecular vibration and rotation are in fact quantised, so that at low temperature the finite energy threshold for the excitation of rotational and vibrational modes may be prohibitive. It is not possible to assess this last effect without knowing the quantum energy steps for rotation and vibration for the molecules concerned. For example, diatomic gas typically has two rotational modes excited at temperatures of a few hundred degrees, so that in this case $C_V = 5\mathcal{R}_*/2\mu$.

Once armed with this relationship between C_V and $\mathcal{R}_*/\mu$ (for a particular atom or molecule) one can thus determine the power law relations between p, V and T (Equations (4.9)–(4.11)) for reversible adiabatic changes.

However, the exponents of these power laws are usually instead expressed in terms of a second specific heat capacity, that at constant pressure, C_p. This is obtained from the perfect gas equation of state by noting that

$$p\,dV + V\,dp = \frac{\mathcal{R}_*}{\mu}dT. \tag{4.12}$$

Thus from (4.4) we have

$$dQ = \left(\frac{d\mathcal{E}}{dT} + \frac{\mathcal{R}_*}{\mu}\right)dT - V\,dp, \tag{4.13}$$

from which it follows that

$$C_p = \left(\frac{\mathrm{d}\mathcal{E}}{\mathrm{d}T} + \frac{\mathcal{R}_*}{\mu} \right), \tag{4.14}$$

and

$$C_p - C_V = \frac{\mathcal{R}_*}{\mu}. \tag{4.15}$$

We now define γ as the ratio of specific heats, i.e.

$$\gamma = \frac{C_p}{C_V}. \tag{4.16}$$

Thus the relations between p, V and T for reversible adiabatic changes may be written:

$$V \propto T^{-1/(\gamma-1)}, \tag{4.17}$$

$$p \propto T^{\gamma/(\gamma-1)}, \tag{4.18}$$

$$p \propto V^{-\gamma}. \tag{4.19}$$

Bearing in mind that the volume V in the above equations refers to that occupied by unit mass of gas, the density is just the reciprocal of V and thus we can write the equation of state of a gas undergoing reversible adiabatic changes in the barotropic form:

$$p = K\rho^{\gamma}, \tag{4.20}$$

where K is constant. Returning now to the first law of thermodynamics (4.3) we note that for a reversible change this can be written $\mathrm{d}Q = T\,\mathrm{d}S$ where S is the entropy per unit mass. A fluid element behaves *adiabatically* if K is constant as the element's properties change, whereas an *isentropic* fluid is one in which all elements have the same value of entropy per unit mass, and in which all therefore share the same value of K.

4.3 Energy equation

We know from the first law of thermodynamics that in the absence of viscous (dissipative) processes, the internal energy in a fluid, $\mathcal{E}$ per kg, is related to the $p\,\mathrm{d}V$ work done, W, and the energy gained from the surroundings, Q, by

$$\frac{\mathrm{D}\mathcal{E}}{\mathrm{D}t} = \frac{\mathrm{d}W}{\mathrm{d}t} + \frac{\mathrm{d}Q}{\mathrm{d}t}, \tag{4.21}$$

and we know that

$$\frac{DW}{Dt} = -p\frac{D(1/\rho)}{Dt} = \frac{p}{\rho^2}\frac{D\rho}{Dt}, \qquad (4.22)$$

so

$$\frac{D\mathcal{E}}{Dt} = \frac{p}{\rho^2}\frac{D\rho}{Dt} - \dot{Q}_{cool}. \qquad (4.23)$$

Here we have defined a cooling function $\dot{Q}_{cool}$ (per kilogram) which is positive if the medium is cooled and negative for external heating.

We now define a total energy *per unit volume*

$$E = \rho\left(\frac{1}{2}u^2 + \Psi + \mathcal{E}\right), \qquad (4.24)$$

where the terms are recognisable as the kinetic energy, potential energy and internal energy. Then

$$\frac{DE}{Dt} = \frac{E}{\rho}\frac{D\rho}{Dt} + \rho\left(\mathbf{u}\cdot\frac{D\mathbf{u}}{Dt} + \frac{D\Psi}{Dt} + \frac{p}{\rho^2}\frac{D\rho}{Dt} - \dot{Q}_{cool}\right). \qquad (4.25)$$

Now we substitute for $\frac{d\rho}{dt}$ and $\frac{d\mathbf{u}}{dt}$ from the continuity and momentum equations, and convert from Lagrangian to Eulerian form:

$$\frac{DE}{Dt} = \frac{\partial E}{\partial t} + \mathbf{u}\cdot\boldsymbol{\nabla}E. \qquad (4.26)$$

For the RHS of Equations (4.25) and (4.26) we use the Lagrangian forms of the continuity and momentum equations, so

$$\frac{E}{\rho}\frac{D\rho}{Dt} = -\frac{E}{\rho}\rho\boldsymbol{\nabla}\cdot\mathbf{u} \quad \text{(continuity)}, \qquad (4.27)$$

$$\rho\mathbf{u}\cdot\frac{D\mathbf{u}}{Dt} = \mathbf{u}\cdot(-\boldsymbol{\nabla}p - \rho\boldsymbol{\nabla}\Psi) \quad \text{(momentum)}, \qquad (4.28)$$

$$\rho\frac{D\Psi}{Dt} = \rho\frac{\partial\Psi}{\partial t} + \rho\mathbf{u}\cdot\boldsymbol{\nabla}\Psi \quad \text{(Eulerian)}, \qquad (4.29)$$

$$\rho\frac{p}{\rho^2}\frac{D\rho}{Dt} = -\rho\frac{p}{\rho^2}\rho\boldsymbol{\nabla}\cdot\mathbf{u} = -p\boldsymbol{\nabla}\cdot\mathbf{u} \quad \text{(continuity)}. \qquad (4.30)$$

As a result, the RHS of Equation (4.25) is

$$= -E\boldsymbol{\nabla}\cdot\mathbf{u} - \mathbf{u}\cdot\boldsymbol{\nabla}p - \rho\mathbf{u}\cdot\boldsymbol{\nabla}\Psi + \rho\frac{\partial\Psi}{\partial t} + \rho\mathbf{u}\cdot\boldsymbol{\nabla}\Psi - p\boldsymbol{\nabla}\cdot\mathbf{u} - \rho\dot{Q}_{cool}$$

$$= -(E+p)\boldsymbol{\nabla}\cdot\mathbf{u} - \mathbf{u}\cdot\boldsymbol{\nabla}p + \rho\frac{\partial\Psi}{\partial t} - \rho\dot{Q}_{cool}. \qquad (4.31)$$

Putting this together with (4.26) and rearranging gives us

$$\frac{\partial E}{\partial t} + \nabla \cdot [(E+p)\mathbf{u}] = -\rho \dot{Q}_{\text{cool}} + \rho \frac{\partial \Psi}{\partial t}. \qquad (4.32)$$

For almost every situation we wish to contemplate $\partial \Psi / \partial t = 0$, i.e. Ψ is a function of position only. Then the only things we have to do are relate E to the quantities we have been dealing with (ρ, T etc.) and sort out an expression for $\dot{Q}_{\text{cool}}$.

Concerning $\dot{Q}_{\text{cool}}$, we have another new variable which we have to determine, for each element in the fluid. So we have apparently merely shifted the problem of closure of the set of equations again, but this time it is more straightforward to see what to do about it. Heating and cooling rates are something we can appreciate (and work out) in an astrophysical context more easily. In the following section we summarize some of the processes which are important in astrophysical environments.

4.4 Energy transport

There are several ways of transporting heat into and out of fluid elements, i.e. conduction, convection and radiation. (Note that if the energy equation is cast in an Eulerian form, then energy may also leave or enter spatially fixed elements as a result of *advection* of heat with the fluid flow.) We shall consider each of these mechanisms for energy transport in turn and briefly discuss their applicability to astrophysical environments. Note that any or all of the processes described involve energy transfer, and so may be applied where they act as a source of energy for the region of interest (and so heating the gas), or a sink of energy from it. However, some radiative processes involve the emission of radiation to the rest of the Universe in a form which interacts little with the material in the region, and so these are generally thought of as cooling processes. Similarly, relativistic particles serve only to inject energy and are regarded as a source of heat.

4.4.1 Cosmic rays

Cosmic rays consist of highly energetic particles, mainly protons, which stream through the Galaxy. When they pass through an interstellar cloud they can ionise atoms and the excess energy in the freed electron ends up as heat. The rate of ionisation per unit volume is proportional to the flux of cosmic rays and the density of material, with rate coefficients which depend on the main constituents of the cloud. For atomic clouds, cosmic ray ionisation of hydrogen atoms injects about $4\,\mathrm{eV}$ $(6 \times 10^{-19}\,\mathrm{J})$ per interaction. Since the cosmic ray flux within a cloud does not depend on the location within the cloud, the cosmic ray heating rate per unit mass is effectively constant.

4.4.2 Conduction

Conduction is the least important astrophysically in almost all contexts. Exceptions to this include the interiors of white dwarfs and the shock fronts induced by the expansion of supernovae into the interstellar medium. Heat conduction is the transfer of heat from warm areas to cooler ones and is a consequence of random particle motions. Particles carry their internal energy with them when they travel from warmer to cooler regions until the point at which they collide with ambient particles. Evidently this process drives a net transfer of energy from hot to cold. The heat flux per unit area is

$$F_{\mathrm{cond}} = -K\nabla T, \qquad (4.33)$$

where K is the thermal conductivity (W m^{-1} K^{-1}). For an ideal gas, K is related to the collision cross-section σ and thermal velocity (hence temperature) according to $K \approx C_V (mkT/3)^{\frac{1}{2}}/\sigma$, where m is the particle mass (see Collins, *The Fundamentals of Stellar Astrophysics*, W. H. Freeman & Co., 1989). The form of this equation is readily explicable in terms of the heuristic picture of conduction outlined above: conduction is favoured by conditions of high temperature (fast particle migration from place to place), low σ (particles travel a long way between collisions and hence transfer heat over large distances) and high C_V (particles carry a lot of thermal energy with them). The rate of change of energy per unit volume due to conduction is simply the divergence of F_{cond} and is therefore $\propto K\nabla^2 T(\mathbf{r}, t)$.

We will see in Chapter 10 that one of the most important roles of conduction is in suppressing thermal instability (i.e. runaway heating and cooling) on small scales, as it tends to equalise the temperature between adjoining regions. Conduction is however suppressed in the presence of magnetic fields, since these inhibit the free streaming of charged particles perpendicular to the field.

4.4.3 Convection

Convection is the transfer of energy by fluid flows which are (usually) set up by gravity in the presence of temperature gradients. It is an important energy transfer mechanism in a number of different types of stars (for example, the cores of massive stars and the envelopes of low mass stars). We shall discuss the process at more length in Chapter 10.

Convection currents arise when a fluid cell is unstable against small displacements, and its motion then carries associated thermal energy, mixing hotter material with cold and effecting a net transfer of energy from hot to cold. Convection operates on scales much larger than individual gas particles and one can therefore consider it as a fluid dynamical process in its own right (for example, one can perform computer simulations that track the circulation of convection currents within stars). On the other hand, one can alternatively consider scales much larger than the scale of convective eddies, and in this case the net effect of convection can be modelled as a heat transfer process. In other words, on these larger scales there is no *net motion* associated with convection but there is still a net energy transfer from hot to cold.

4.4.4 Radiation

Energy carried by photons tends to dominate in many astrophysical contexts. Radiative energy loss is a complex problem which is most easily understood at a qualitative level if one considers two limiting cases. In one (optically thick) limit, the emitted photons are reabsorbed or scattered locally and have to escape to infinity only after diffusing outwards through the medium. In the optically thin limit, by contrast, photons are able to escape to infinity as the overlying material is nearly transparent to the emitted radiation.

The optically thick limit may be understood as a diffusion problem, analogous to the process of conduction described above, the difference here being that the particles that travel between hotter and colder regions are now photons instead of atoms or molecules. A further difference is that photon number is not conserved – hotter regions emit more photons than cool regions and with higher average photon energy so that the net effect is again one of heat transfer from hot to cold. This process continues as the photons diffuse through successively cooler regions until a region is reached, known as the photosphere, where they (on average) have their last interaction with the atoms in the fluid before escaping to infinity. The fact that the photons have interacted many times with the fluid on their way out means that their energy spectrum is in a state of thermodynamic equilibrium with the fluid at every point in the fluid until the photosphere is reached. The observer at infinity therefore sees only a black body spectrum characteristic of the temperature of the photosphere. This (in grossly oversimplified form) is the case for stars – the spectrum tells the observer at infinity about the temperature of the photosphere, but the multiple interactions with the fluid particles have erased any memory of the physical processes involved in individual photon emission events. In practical terms, for those studying the structure of stars, one can use the radiative diffusion approximation to calculate the net heat flux by radiation at any point in the star provided one knows its temperature and density structure and the way that opacity depends on temperature and density. (The reason that the opacity is important is that it affects the photons' mean free path and thus, as in the case of conduction discussed above, regulates the distances over which photons can transfer energy.)

The optically thin limit (i.e. the case of radiative loss to infinity) is simple in principle, but a little messy in practice. It consists simply of adding up the contributions from all the relevant processes.

The sorts of loss processes we might consider are:

(i) Energy loss by recombination is usually only a minor contributor. For hydrogen, the energy lost by free electrons per unit volume per unit time is

$$L_{\text{recomb}} = n_e n_p k T \beta (H^0, T), \qquad (4.34)$$

where n_e is the electron number density (particles per cubic metre), n_p the proton number density and $kT\beta(H^0, T)$ is the product of the particle kinetic energy, recombination cross-section and velocity all averaged over particle velocities. Since, in a Maxwellian velocity distribution, the particle velocity scales as the inverse square root of the particle mass, it is the electron (rather than the proton) velocity distribution that is relevant to this calculation.

(ii) Energy loss by free–free radiation (sometimes called bremsstrahlung). This occurs when an electron is accelerated as it passes close to a charged particle of charge Z, and results in an energy loss

$$L_{\text{ff}} = 1.42 \times 10^{-40} Z^2 \sqrt{T} g_{\text{ff}} n_e n_p \quad \text{W m}^{-3}, \qquad (4.35)$$

where g_{ff} is weakly temperature dependent, and may be taken to be unity. It is not usually the dominant energy loss mechanism, but does occur in any region where there are free electrons and ions.

(iii) Collisionally excited atomic line radiation. For many applications this one wins, and is one of the most complicated to deal with.

The energy loss is dominated by electron collisions of atoms in the ground state which give rise to excitation to a low-lying energy level, and the atom then returns to the ground state by emitting a photon (of energy χ) which takes away the excitation energy. Thus electron kinetic energy is effectively radiated away. The cooling rate per unit volume per unit time for line radiation from a given energy level for a particular species is

$$L_C = n_{\text{ion}} n_e e^{-\chi/kT} \chi \times \frac{8.6 \times 10^{-12}}{\sqrt{T}} \times \omega \quad \text{W m}^{-3}. \tag{4.36}$$

This expression contains both ion and electron densities, since collisions are involved in the excitation process, and the Boltzmann factor $(e^{-\chi/kT})$ represents the probability of excitation of a given transition at temperature T. The next term (χ) is the energy released per radiative de-excitation and the following terms are temperature-dependent rate coefficients obtained by averaging over a Maxwellian particle velocity distribution (the term ω depends on the statistical weights of the energy levels involved in the transition). In relatively cold gas clouds (e.g. 10^4 K) there is not enough energy to excite hydrogen, so common ions like O^+, O^{++} and N^+ which have low-lying excited levels provide the bulk of the cooling. Much of the light we see from gaseous nebulae, such as that in Figure 4.1, is line radiation arising in this way.

One thing that is evident from this discussion is that the rates depend on the densities of ions, not atoms, and we thus still have to determine

Fig. 4.1. The Helix planetary nebula, the visible light from which is dominated by recombination lines of hydrogen and collisionally excited lines of oxygen and nitrogen. (Space Telescope Science Institute)

the densities of each relevant ionic species. These can be derived by considering equilibrium between recombination and ionisation processes. The former rate is closely related to the cooling rate due to recombination (Equation (4.34)), whereas ionisation may be either collisional or radiative. Collisional ionisation is relevant at high temperatures and the rate is then of the form

$$n_{H^0} n_e e^{-\chi/kT} \alpha$$

for ionisation of neutral hydrogen. However, in many cases ionisation occurs when a photon from e.g. a nearby star has enough energy to separate an electron from an atom. Then the ionisation rate is

$$n_{H^0} \int_{\nu_0}^{\infty} \frac{F_\nu}{h\nu} a_\nu \, d\nu,$$

where ν is the radiation frequency, F_ν the photon energy flux per hertz (and so $F_\nu/h\nu$ the number flux of the photons), a_ν an absorption coefficient which depends on the species being considered, and the integral goes from ν_0, the frequency corresponding to the lowest energy required to ionise the atom. Of course this process also heats the medium so provides another energy input term in the thermal equation. Finally, the whole system is closed by assuming element abundances, and the requirement that the sum of all ionisation stages is the total density of a particular element.

(iv) For cool interstellar regions, e.g. where star formation may occur with temperatures of a few 100 K or less, collisions between molecules can provide significant cooling. When two molecules collide, some of their kinetic energy can be transferred to rotational or internal vibration, or into exciting one or more electrons in the molecules. Where the energy loss mechanism is by photon emission then this acts as a cooling mechanism in the same way as for the atomic processes described above. As with the atomic case, the cooling rate depends on density, abundances of various molecules and the temperature, and involves summing over a number of contributors. A range of molecules can be important, with CO and O_2 dominating at densities $< 10^{12} \, H_2$ molecules/m^3 at temperatures ~ 20 K, and H_2O and other species gaining in importance at higher densities.

Such regions also tend to contain significant amounts of dust, so there are the added components of thermal radiation by dust grains at far infrared wavelengths as an energy loss process, and absorption of trapped hydrogen Ly α photons as one of a number of mechanisms for heat gain. The dust grains exchange heat with the gas phase by processes such as photoelectric ejection and electron recombination. These will not be explored further here, but left to one of several good sources of material on interstellar medium physics, e.g. Tielens, *The Physics and Chemistry of the Interstellar Medium* (Cambridge University Press, 2005).

The details of such calculations are far beyond the scope of this book (D. E. Osterbrock, *Astrophysics of Gaseous Nebulae and Active Galactic Nuclei*, University Science Books, 1989 gives an in-depth treatment). Generally, for astrophysical fluid dynamics at this level, we assume either a barotropic equation of state (see Section 4.3) or else a simple cooling law such as the cooling rate for free-free emission from an optically thin, completely ionised nebula (Equation (4.35)).

Each of the radiative loss processes above is reversible, so can, in the right circumstances, act as a heat source for a region. So, for example, a gaseous region near a hot star will gain heat through absorption of continuous radiation by the inverse process to that which leads to heat loss (see Equation (4.34)). The energy gained depends on such things as the energy spectrum of the incoming radiation, and the frequency-dependent optical depth of the region between the star and the region of interest. This adds a further level of complexity, and again the reader is referred to Osterbrock's book. Energy gain from absorption of continuum radiation by line transitions is one thing which can usually be ignored.

4.5 The form of $\dot{Q}_{\mathrm{cool}}$

Since the cooling rate per unit volume in the optically thin case commonly scales as the square of the density, it then follows that the cooling rate per unit mass scales linearly with density, and so we may parameterise this over a limited temperature range by $A\rho T^{\alpha}$, where A and α are constants. In dense clouds the photon flux is essentially zero, so cosmic rays provide almost all the heating, and so the heating rate per unit mass is approximately constant (see Section 4.4.1). Hence we will find the parameterisation

$$\dot{Q}_{\mathrm{cool}} = A\rho T^{\alpha} - H \tag{4.37}$$

useful. We use a net cooling rate of this form in Chapter 10 when considering thermal instability.

We conclude this discussion of thermal processes by noting the strong (quadratic) density dependence of the optically thin cooling rate per unit volume from a variety of processes (usually from equations like (4.36) with contributions summed over a variety of ions). Thus we should bear in mind that there are circumstances where density enhancements in low density media can precipitate rapid cooling. This is an issue that we will return to when we consider the evolution of supernova blast waves in the interstellar medium in Chapter 8.

Chapter 5
Hydrostatic equilibrium

We are now ready to solve the fluid equations and will start with the simplest case, that of hydrostatic equilibrium.

5.1 Basic equations

Hydrostatic equilibrium implies that $u = 0$ everywhere ('static') and that $\partial/\partial t = 0$ ('equilibrium'). The continuity equation is trivially satisfied and in this chapter we will consider barotropic equations of state (see Section 4.2) so that we can dispense with the energy equation. The only equation to be solved is therefore the momentum equation in which the only non-zero terms are now gravity and pressure, which must therefore balance. Thus

$$\frac{1}{\rho}\nabla p = \mathbf{g}, \qquad (5.1)$$

or, equivalently,

$$\frac{1}{\rho}\nabla p = -\nabla\Psi. \qquad (5.2)$$

If we know the barotropic equation of state $(p = p(\rho))$, we can then use Poisson's equation $(\nabla^2\Psi = 4\pi G\rho)$ to solve for the density distribution $\rho(\mathbf{r})$ corresponding to hydrostatic equilibrium everywhere. Naturally, this will also provide solutions for p and Ψ everywhere as well.

We will now consider a few examples where the symmetry of the problem allows Gauss's theorem to be applied straightforwardly.

5.2 The isothermal slab

We consider an infinite (in x and y) static isothermal slab, symmetric about $z = 0$, supported by gas pressure and under its own self-gravity. No other forces are acting. An example of the sort of situation where this might arise in an astronomical context is if two clouds collide and generate a shocked slab of gas between them. Of course we are never in reality dealing with infinite systems but for our purposes here we just require that the slab diameter is very large compared with the slab thickness and hence that we can consider all gravitational forces to act in the z direction.

Since we are dealing with an isothermal ideal gas, the equation of state $p = \frac{\mathcal{R}_*}{\mu} \rho T$ may be written in the form $p = A\rho$, where A is a constant. The geometry is an infinite slab in x, y, so $\nabla = \partial/\partial z$, and ρ and Ψ are functions of z only. Substituting in the hydrostatic equilibrium equation above gives

$$A \frac{1}{\rho} \nabla \rho = -\nabla \Psi. \tag{5.3}$$

This becomes

$$A \frac{d}{dz} \ln \rho = -\frac{d\Psi}{dz} \tag{5.4}$$

$$\Rightarrow \Psi = -A \ln \left(\frac{\rho}{\rho_0} \right) + \Psi_0, \tag{5.5}$$

where $\rho_0 = \rho|_{z=0}$ and $\Psi_0 = \Psi|_{z=0}$. Hence

$$\rho = \rho_0 e^{-\frac{(\Psi - \Psi_0)}{A}}. \tag{5.6}$$

Now we may use Poisson's equation to find the z dependence of Ψ:

$$\frac{d^2\Psi}{dz^2} = 4\pi G \rho_0 e^{-\frac{(\Psi - \Psi_0)}{A}}. \tag{5.7}$$

We change variables to $\chi = -(\Psi - \Psi_0)/A$ and $Z = (2\pi G \rho_0/A)^{\frac{1}{2}} z$ (the factor 2 instead of 4 within the square root is not obvious, and nor does it matter much – it just simplifies the coefficients a bit later). Then

$$\frac{d^2\chi}{dZ^2} = -2e^\chi, \tag{5.8}$$

with boundary conditions $\chi = 0$ at $Z = 0$ and, if we assume that $z = 0$ is the plane of symmetry, $\frac{d\chi}{dZ} = 0$ at $Z = 0$.

Multiplying the equation by $d\chi/dZ$, we obtain

$$\frac{d\chi}{dZ}\frac{d^2\chi}{dZ^2} = -2\frac{d\chi}{dZ}e^\chi \tag{5.9}$$

$$\Rightarrow \frac{1}{2}\frac{d}{dZ}\left[\left(\frac{d\chi}{dZ}\right)^2\right] = -2\frac{d}{dZ}[e^\chi] \tag{5.10}$$

$$\Rightarrow \left(\frac{d\chi}{dZ}\right)^2 = c_1 - 4e^\chi. \tag{5.11}$$

The boundary conditions at $Z = 0 \Rightarrow c_1 = 4$, and so

$$\frac{d\chi}{dZ} = 2(1-e^\chi)^{\frac{1}{2}}. \tag{5.12}$$

All we have to do now is integrate this, which means changing variables until we see something we recognise. For

$$\int \frac{d\chi}{(1-e^\chi)^{\frac{1}{2}}}, \tag{5.13}$$

we try $s^2 = e^\chi$ since the standard integrals with a square root of a constant plus a variable in the denominator usually have $\sqrt{(a^2 \pm s^2)}$ there. This gives

$$\int \frac{2ds}{s(1-s^2)^{\frac{1}{2}}}. \tag{5.14}$$

Now we substitute $s = \sin\theta$, so $ds = \cos\theta\, d\theta$, and so we have

$$2\int \frac{d\theta}{\sin\theta}. \tag{5.15}$$

The manipulation is now a bit more standard. Substituting $t = \tan\frac{\theta}{2}$, then $dt = \frac{1}{2}(1+t^2)\, d\theta$ and $\sin\theta = 2t/(1+t^2)$, and the integral becomes

$$2\int \frac{dt}{t}. \tag{5.16}$$

Therefore

$$2\ln t = 2Z + c_2, \tag{5.17}$$

where c_2 is a constant of integration.

The boundary condition $\chi = 0$ when $Z = 0$ leads to $s = 1$ at $Z = 0$. This in turn implies $\theta = \pi/2$, $t = 1$, and $c_2 = 0$. Hence

$$t = e^Z. \tag{5.18}$$

$s = 2t/(1 + t^2)$, and so

$$s = e^{\frac{\chi}{2}} = \frac{2e^Z}{(1 + e^{2Z})} = \frac{1}{\cosh Z}. \tag{5.19}$$

Substituting back in the variables we had at the start, we find

$$\Psi - \Psi_0 = 2A \ln \cosh \left(\sqrt{\frac{2\pi G \rho_0}{A}} z \right). \tag{5.20}$$

Thus we have determined the density profile for an isothermal slab:

$$\rho = \frac{\rho_0}{\cosh^2 \left(\sqrt{\frac{2\pi G \rho_0}{A}} z \right)}. \tag{5.21}$$

Curiously enough, after such exhausting substitutions, the solution $\rho \propto \operatorname{sech}^2(az)$ for some constant a is rather simple. The reader is directed to Exercise 16 where the nature of the solution is explored in more detail.

5.3 An isothermal atmosphere with constant g

We can treat the Earth's atmosphere, for example, as a plane (since g is a constant, i.e. $1/r^2$ effects are small near to the surface of the Earth) and we may also treat it as approximately isothermal. Again we have

$$A \frac{1}{\rho} \nabla \rho = -\nabla \Psi = -g \tag{5.22}$$

$$\Rightarrow \ln \rho = -\frac{gz}{A} + \text{constant}, \tag{5.23}$$

i.e. we have an exponential atmosphere.

$$\rho = \rho_0 \exp \left(\frac{-\mu g}{\mathcal{R}_* T} z \right). \tag{5.24}$$

For the Earth's atmosphere, $T \sim 300\,\text{K}$, $\mu \sim 28$, and one finds an e-folding height that is about $9\,\text{km}$. The top of Mauna Kea in Hawaii is the site of many of the world's most powerful telescopes, owing to the superior seeing conditions (and also the low water vapour content) at the summit. However, since the altitude is $4.2\,\text{km}$, astronomers observing there have to contend with working, thinking and staying awake overnight with atmospheric density and pressure $\sim 60\%$ of the values near sea level.

5.4 Stars as self-gravitating polytropes

We assume we have a spherical system in hydrostatic equilibrium i.e. there is no rotation which would break spherical symmetry. (Note that this analysis would also be applicable to a star rotating as a solid body, in which case centrifugal effects can be included as an effective potential: in this case, the radial coordinate r would be replaced by a coordinate perpendicular to equipotential surfaces.) We have, as always,

$$\nabla p = -\rho \, \nabla \Psi. \tag{5.25}$$

In spherical polars this is, unsurprisingly,

$$\frac{\mathrm{d}p}{\mathrm{d}r} = -\rho \frac{\mathrm{d}\Psi}{\mathrm{d}r}. \tag{5.26}$$

Since $\rho > 0$ within the star, p is a monotonic function of Ψ and we can write $p = p(\Psi)$.

$$\frac{\mathrm{d}p}{\mathrm{d}r} = \frac{\mathrm{d}p}{\mathrm{d}\Psi}\frac{\mathrm{d}\Psi}{\mathrm{d}r} = -\rho \frac{\mathrm{d}\Psi}{\mathrm{d}r}, \tag{5.27}$$

so

$$\rho = -\frac{\mathrm{d}p}{\mathrm{d}\Psi}, \tag{5.28}$$

$$\Rightarrow \rho = \rho(\Psi). \tag{5.29}$$

Since $p = p(\Psi)$ and $\rho = \rho(\Psi)$, it follows that $\rho = \rho(p)$ which is the definition of a *barotropic* equation of state. In practice we can often additionally assume that ρ is a monotonic function of p and in this case can usefully parametrise the $p - \rho$ relation as a *polytrope*:

$$p = K\rho^{1+\frac{1}{n}}, \tag{5.30}$$

where the parameter n is known as the polytropic index. This is purely for convenience, and in the knowledge that over a limited range one can always fit the p–ρ relation as a power law. In fact it turns out to be a reasonably good approximation to use a single power law (i.e. single n) throughout the entire interior of some stars, so that a polytropic approach is a reasonable one. Note that the adiabatic constant γ is *not* equal to the polytropic power law $1 + \frac{1}{n}$ in general: a star that obeys $p \propto \rho^{\gamma}$ has the special property that it is isentropic, whereas a polytropic star is in general not isentropic but just has a barotropic equation of state that can be approximated as a power law.

To solve for the internal structure of a (polytropic) star, one substitutes the polytropic equation of state into the equation of hydrostatic equilibrium:

$$-\nabla\Psi = \frac{1}{\rho}\nabla\left(K\rho^{1+\frac{1}{n}}\right) = (n+1)\nabla\left(K\rho^{\frac{1}{n}}\right). \qquad (5.31)$$

Therefore

$$\rho = \left(\frac{\Psi_{\mathrm{T}} - \Psi}{(n+1)K}\right)^{n}, \qquad (5.32)$$

where we define Ψ_{T} as the value of the potential where $\rho = 0$ (and the equations still apply), i.e. the surface of the star.

Now we use Poisson's equation:

$$\nabla^2\Psi = 4\pi G\rho = 4\pi G\left(\frac{\Psi_{\mathrm{T}} - \Psi}{(n+1)K}\right)^{n}. \qquad (5.33)$$

Defining ρ_{c} and Ψ_{c} as the values of the density and the potential at the centre of the star, we have

$$\rho = \rho_{\mathrm{c}}\left(\frac{\Psi_{\mathrm{T}} - \Psi}{\Psi_{\mathrm{T}} - \Psi_{\mathrm{c}}}\right)^{n} \qquad (5.34)$$

and

$$\nabla^2\Psi = 4\pi G\rho_{\mathrm{c}}\left(\frac{\Psi_{\mathrm{T}} - \Psi}{\Psi_{\mathrm{T}} - \Psi_{\mathrm{c}}}\right)^{n}. \qquad (5.35)$$

Setting $\theta = \left(\frac{\Psi_{\mathrm{T}} - \Psi}{\Psi_{\mathrm{T}} - \Psi_{\mathrm{c}}}\right)$, so $\Psi = -(\Psi_{\mathrm{T}} - \Psi_{\mathrm{c}})\theta + \Psi_{\mathrm{T}}$,

$$\Rightarrow \nabla^2\theta = -\frac{4\pi G\rho_{\mathrm{c}}}{\Psi_{\mathrm{T}} - \Psi_{\mathrm{c}}}\theta^{n}. \qquad (5.36)$$

Since we have spherical polars,

$$\nabla^2\theta = \frac{1}{r^2}\frac{\mathrm{d}}{\mathrm{d}r}\left(r^2\frac{\mathrm{d}\theta}{\mathrm{d}r}\right),$$

we can change variables to

$$\xi = \left(\frac{4\pi G\rho_c}{\Psi_T - \Psi_c} \right)^{\frac{1}{2}} r, \tag{5.37}$$

and the equation then becomes

$$\frac{1}{\xi^2} \frac{d}{d\xi} \left(\xi^2 \frac{d\theta}{d\xi} \right) = -\theta^n. \tag{5.38}$$

This is the 'Lane–Emden equation of index n'. This has to be integrated (numerically for a general n) subject to two boundary conditions:

- $\theta = 1$ at $\xi = 0$ (from the definition of θ),
- $\frac{d\theta}{d\xi} = 0$ at $\xi = 0$. This implies zero force at $\xi = 0$, which is true for gas spheres because the enclosed mass $\to 0$ as $\xi \to 0$. It is not true if there is a point mass in the middle of the sphere.

This equation can be solved analytically for particular values of n (0, 1 and 5) but not for general n.

5.5 Solutions for the Lane–Emden equation

5.5.1 Solution for $n = 0$

The Lane–Emden equation for $n = 0$ is

$$\frac{1}{\xi^2} \frac{d}{d\xi} \left(\xi^2 \frac{d\theta}{d\xi} \right) = -1. \tag{5.39}$$

So, integrating once,

$$\xi^2 \frac{d\theta}{d\xi} = -\frac{1}{3}\xi^3 - C, \tag{5.40}$$

where C is a constant. Therefore

$$\theta = D + \frac{C}{\xi} - \frac{1}{6}\xi^2, \tag{5.41}$$

where D is another constant.

This solution has a singularity at the origin, and $\theta \to \frac{C}{\xi}$ as $\xi \to 0$. However, the boundary conditions at $\xi = 0$ are $\theta = 1$ and $\frac{d\theta}{d\xi} = 0$, and the only way to satisfy these is for $C = 0$ and $D = 1$. So the solution for $n = 0$ is

$$\theta_0 = 1 - \frac{1}{6}\xi^2. \tag{5.42}$$

The surface of the polytrope is where the function θ is zero, i.e. at $\xi_1 = \sqrt{6}$. (The subscript 0 is just to highlight that this is the solution only for $n = 0$.)

5.5.2 Solution for $n = 1$

The equation for $n = 1$ is

$$\frac{1}{\xi^2}\frac{d}{d\xi}\left(\xi^2\frac{d\theta}{d\xi}\right) = -\theta. \tag{5.43}$$

This is more tractable if we set $\theta = \chi/\xi$ and solve for χ. Then the Lane–Emden equation is, in general,

$$\frac{d^2\chi}{d\xi^2} = -\frac{\chi^n}{\xi^{n-1}}, \tag{5.44}$$

and in this particular case

$$\frac{d^2\chi}{d\xi^2} = -\chi. \tag{5.45}$$

This is very familiar, with solution

$$\chi = A\sin(\xi + B), \tag{5.46}$$

where A and B are constants of integration. Therefore

$$\theta = \frac{A\sin(\xi + B)}{\xi}. \tag{5.47}$$

The boundary conditions then require that $A = 1$ and $B = 0$, so the solution for $n = 1$ is

$$\theta_1 = \frac{\sin\xi}{\xi}. \tag{5.48}$$

This has its first zero at $\xi = \pi$, and is monotonically decreasing over the range $(0, \pi)$.

5.5.3 Solution for $n = 5$

$$\frac{1}{\xi^2}\frac{d}{d\xi}\left(\xi^2\frac{d\theta}{d\xi}\right) = -\theta^5 \tag{5.49}$$

is not something that suggests there is an analytic solution. However, it does have one.

In general, if one makes the transformation $x = 1/\xi$, the Lane–Emden equation becomes

$$x^4\frac{d^2\theta}{dx^2} = -\theta^n. \tag{5.50}$$

Then setting $t = \ln x$ and

$$\theta = \left[\frac{2(n-3)}{(n-1)^2}\right]^{\frac{1}{(n-1)}} x^{\frac{2}{n-1}} z, \tag{5.51}$$

the equation in z looks like

$$\frac{d^2z}{dt^2} + \frac{5-n}{n-1}\frac{dz}{dt} - \frac{2(n-3)}{(n-1)^2}z(1-z^{n-1}) = 0. \tag{5.52}$$

In the case where $n = 5$ this simplifies to

$$\frac{d^2z}{dt^2} = \frac{1}{4}z(1-z^4), \tag{5.53}$$

where $\frac{1}{x} = \xi = e^{-t}$ and $\theta = \left(\frac{x}{2}\right)^{\frac{1}{2}} z = \left(\frac{1}{2}e^t\right)^{\frac{1}{2}} z$. Multiplying both sides of the equation by $\frac{dz}{dt}$, then

$$\frac{1}{2}\frac{d}{dt}\left[\left(\frac{dz}{dt}\right)^2\right] = \frac{1}{4}z(1-z^4)\frac{dz}{dt}, \tag{5.54}$$

so

$$\frac{1}{2}\left(\frac{dz}{dt}\right)^2 = \frac{1}{8}z^2 - \frac{1}{24}z^6 + D, \tag{5.55}$$

where D is a constant. The boundary conditions imply $D = 0$, so one now has to solve

$$\frac{dz}{z(1-\frac{1}{3}z^4)^{\frac{1}{2}}} = -\frac{1}{2}dt, \tag{5.56}$$

with a choice of sign for the square root which allows $t \to \infty$.

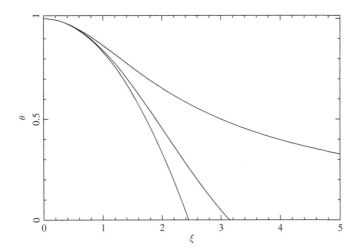

Fig. 5.1. Solutions to the
Lane–Emden equation for
index $n = 0$, 1 and 5. The
radius at which $\theta = 0$
increases with increasing n.

To solve this, one makes the substitution $\frac{1}{3}z^4 = \sin^2 \phi$, and then finds that $\tan \frac{1}{2}\phi = Ce^{-t}$. Putting this back in with the boundary conditions then yields

$$\theta_5 = \frac{1}{(1 + \frac{1}{3}\xi^2)^{\frac{1}{2}}}. \tag{5.57}$$

Note that for this equilibrium solution, $\theta \to 0$ only as $\xi \to \infty$, which means that the equilibrium configuration extends to infinity.

(As an aside: The equation expressed in the form given by (5.52) is easily integrated if $n = 3$ also, since the last term then disappears. However, it is a solution of no interest, since from (5.51) we find that θ is identically zero, so the density is zero everywhere.)

5.6 The case of $n = \infty$

The formalism above does not include the special case of the isothermal gas sphere, where $p = K\rho$, since that corresponds to $n = \infty$. One can derive a similar equation of equilibrium, which is

$$K\frac{1}{r^2}\frac{\mathrm{d}}{\mathrm{d}r}\left(\frac{r^2}{\rho}\frac{\mathrm{d}\rho}{\mathrm{d}r}\right) = -4\pi G\rho. \tag{5.58}$$

Then substituting $\rho = \rho_c e^{-\psi}$, $r = \left[\frac{K}{4\pi G\rho_c}\right]^{\frac{1}{2}} \xi \equiv a\xi$ gives

$$\frac{1}{\xi^2}\frac{\mathrm{d}}{\mathrm{d}\xi}\left(\xi^2\frac{\mathrm{d}\psi}{\mathrm{d}\xi}\right) = e^{-\psi}. \tag{5.59}$$

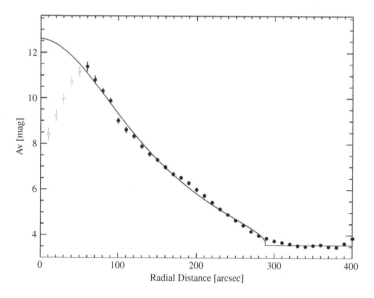

Fig. 5.2. The radial density profile of a dense core of the molecular cloud Coalsack G2, as measured from the extinction levels of background stars. The data in the radial range 55 to 400 arcsec are well fitted by a Bonnor–Ebert sphere truncated at $\xi = 5.8$. (From Lada *et al.*, *Astrophysical Journal* **610**, 303, 2004)

ρ_c is chosen as the central density, so $\psi = 0$ at $\xi = 0$, and the other condition is that $\frac{d\psi}{d\xi} = 0$ at $\xi = 0$. At large radii the solution of (5.59) subject to these boundary conditions tends to $\rho \propto r^{-2}$, from which one deduces that for an isothermal sphere of self-gravitating gas of infinite radius, the mass tends to infinity.

Isothermal spheres of finite mass therefore need to be truncated at some radius, and may exist in hydrostatic equilibrium if they are embedded in an external medium of appropriate pressure. The density profile of such ('Bonnor–Ebert') spheres depends on the value of ξ at which they are truncated. For truncation at small radii the density profile is relatively flat, whereas for truncation at large radii the density profile approaches $\rho \propto r^{-2}$ over a large range of radii. Although the structure of Bonnor–Ebert spheres is a matter of academic interest mainly, it is notable that in recent years the structure of dense gas cores in molecular clouds has been shown to be well fitted by such spheres (see Figure 5.2).

5.7 Scaling relations

There are various conditions under which stars behave like polytropes amongst which is the case of fully convective stars where the pressure–density relation is very close to the adiabatic one. (The reason for this is explained in Chapter 10 where we examine convection in more detail.) For monatomic gases, $\gamma = 5/3$ (see Section 4.1) and thus $p = K\rho^{\frac{5}{3}}$, implying $n = 3/2$. For such cases we have to solve the equations numerically, but we can make some progress with the general properties of such stars through so-called scaling relations. The basic principle is that one treats all the stars characterised by a given polytropic index n as belonging to a 'family', distinguished from each other by the single parameter of the value of the central density, ρ_c (we shall assume for now that all stars in a particular family also share the same value of polytropic constant K). If we can find how quantities such as the mass of the star and its radius vary as functions of ρ_c, we can eliminate ρ_c and discover the relationship between masses and radii for such stars.

The reason that we can do this is that *all* stars with a given value of n have the same $\theta(\xi)$. (To see why this is the case, trace through the derivation of (5.38) and see that this equation, which does not involve ρ_c, is true for all ρ_c.) The value of ρ_c does however determine the mapping between ξ and r and θ and ρ. From (5.32) we obtain

$$\Psi_T - \Psi_c = K(n+1)\rho_c^{\frac{1}{n}},$$ (5.60)

and so (from (5.37)) we have

$$\xi = \left(\frac{4\pi G \rho_c^{1-\frac{1}{n}}}{K(n+1)} \right)^{\frac{1}{2}} r.$$ (5.61)

The mapping between θ and ρ is given by $\rho = \rho_c \theta^n$, so for a given ρ_c, knowing $\theta(\xi)$ tells you $\rho(r)$.

Now, for any polytrope of a given n, there is a particular value of ξ (ξ_{max}) at which $\theta = 0$. The total mass of the star is then

$$M = \int_0^{r_{max}} 4\pi \rho r^2 \, dr$$ (5.62)

so

$$M = 4\pi \rho_c \left(\frac{4\pi G \rho_c^{1-\frac{1}{n}}}{K(n+1)} \right)^{-\frac{3}{2}} \int_0^{\xi_{max}} \theta^n \xi^2 \, d\xi.$$ (5.63)

The integral is now some quantity which is the same for all ρ_c, and so we have that the mass of the star

$$M \propto \rho_c^{\frac{1}{2}(\frac{3}{n}-1)}.$$ (5.64)

Also the radius of the star to the surface

$$r_{max} = \xi_{max} \left(\frac{4\pi G \rho_c^{1-\frac{1}{n}}}{K(n+1)} \right)^{-\frac{1}{2}} \propto \rho_c^{\frac{1}{2}(\frac{1}{n}-1)}. \tag{5.65}$$

Therefore we have a mass–radius relation for polytropic stars:

$$M \propto R^{\frac{3-n}{1-n}}. \tag{5.66}$$

For an adiabatic star we have $1 + \frac{1}{n} = \gamma$, and for a monatomic gas $\gamma = 5/3$ so $n = 3/2$. Hence

$$M \propto R^{-3}. \tag{5.67}$$

So the more massive such a star is, the smaller is its radius. Contrast this with an incompressible star, where ρ is independent of p (since ρ is a constant) so $n = 0$, and $M \propto R^3$.

It seems counter-intuitive that a more massive star should be smaller, but if one examines the derivation one can satisfy oneself that the physical reason for this is simply that the stronger gravity of a massive star squeezes the star into a smaller radius. However satisfied one is with this explanation, observations say otherwise – in real stars $M \propto R$ is a reasonable approximation! So what is wrong?

The most questionable part of what we have done so far is to assume that all stars share a common polytropic constant K, for why, after all, should this be the case? If we combine the polytropic equation of state (5.30) with the ideal gas equation of state we see that the temperature at the centre, T_c, is given by

$$T_c = \frac{\mu K}{\mathcal{R}_*} \rho_c^{\frac{1}{n}}. \tag{5.68}$$

Therefore, if K is the same for all stars then the central temperature varies with ρ_c for different members of the 'family'. However, in reality, nuclear reactions in the core of the star keep the value of the central temperature rather similar in stars of different masses. The constraint $T_c = $ constant then requires that K must vary, i.e.

$$K \propto \rho_c^{-\frac{1}{n}}. \tag{5.69}$$

Then

$$M = 4\pi\rho_c \left(\frac{K(n+1)}{4\pi G \rho_c^{1-\frac{1}{n}}} \right)^{\frac{3}{2}} \int_0^{\xi_{max}} \theta^n \xi^2 \, d\xi \propto \rho_c^{-\frac{1}{2}} \tag{5.70}$$

and

$$R = \left(\frac{K(n+1)}{4\pi G \rho_c^{1-\frac{1}{n}}} \right)^{\frac{1}{2}} \xi_{max} \propto \rho_c^{-\frac{1}{2}} \tag{5.71}$$

and we now obtain the observed relation:

$$M \propto R. \tag{5.72}$$

This would seem to suggest that the scaling relations derived at constant K (e.g. Equation (5.66)) are of little use in astrophysics. However, although there is no reason why different stars should share a common value of K, the case is different if one instead considers the effect of adding mass to a given star. Initially, the star may re-adjust to the addition of mass by attaining a new hydrostatic equilibrium but on a timescale sufficiently short that this happens adiabatically (i.e. no heat is at first exchanged with the surroundings). If the star is isentropic in the first place (as would be expected for a fully convective star), the new structure will also then be isentropic, i.e. share the same K. Therefore scaling relations such as (5.66) are likely to be relevant to stars' re-adjustment when mass is added to them.

Before considering the astronomical contexts in which this is important, we must see whether it is plausible that a star should adjust to a new hydrostatic equilibrium before it has had a chance to exchange significant heat energy with the surroundings. The timescale on which the star adjusts to a new hydrostatic equilibrium ($\nabla p = \rho \mathbf{g}$) is roughly the timescale for a sound wave to propagate across the star (see Chapter 6) which, for the Sun, is less than a day. On the other hand the timescale on which the star can lose significant energy to the surroundings is the thermal timescale:

$$t_{th} \sim \frac{\text{energy content of the star}}{\text{luminosity}} \sim \frac{GM^2}{RL}. \tag{5.73}$$

(In the second equality we have used the virial theorem to equate, as an order of magnitude estimate, the thermal energy content of the star to its potential energy; see Chapter 3.)

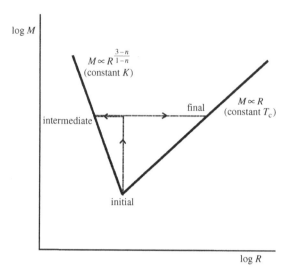

Fig. 5.3. The evolution in the (M, R) plane for a polytrope which is subject to a one-off addition of mass: on a sound-crossing timescale t_{th} the star joins the appropriate place on the constant K scaling relation, but later, on a thermal timescale t_{th} expands to lie on the scaling relation at constant T_c.

Putting in the numbers for the Sun ($1\ M_\odot = 2 \times 10^{33}\ \text{kg}$; $G = 6.67 \times 10^{-11}\ \text{m}^3\ \text{kg}^{-1}\ \text{s}^{-2}$; $1\ R_\odot = 7 \times 10^8\ \text{m}$; $1\ L_\odot = 4 \times 10^{26}\ \text{W}$) gives $t_{th} \sim 10^{15}\ \text{s}$, or $\sim 3 \times 10^7$ years. This is so comfortably larger than the timescale for re-adjustment to hydrostatic equilibrium quoted above that we can safely assume that the immediate response to mass loss or gain is a star that is in a new hydrostatic equilibrium but can remain out of thermal equilibrium for an extended period. If the mass was added in a one-off fashion then eventually, after t_{th} evaluated above, the new structure would 'notice' that it was not in thermal equilibrium and relax to a new thermal equilibrium structure (with different K but close to the original T_c). These ideas are expressed in Figure 5.3.

5.8 Examples of astrophysical interest

(i) Consider the case of a spherical star rotating with angular velocity Ω, on which is dropped a small amount of non-rotating gas. How does Ω evolve thereafter?

We have to conserve angular momentum $J \propto MR^2\Omega$, so if $\Omega \rightarrow \Omega + \Delta\Omega$,

$$MR^2\Delta\Omega + \Omega\Delta(MR^2) = 0, \tag{5.74}$$

or

$$\frac{\Delta\Omega}{\Omega} = -\frac{\Delta(MR^2)}{MR^2}. \tag{5.75}$$

Now

$$R \propto M^{\frac{1-n}{3-n}},$$ (5.76)

so

$$\frac{\Delta\Omega}{\Omega} \propto -\Delta\left(M^{\frac{5-3n}{3-n}}\right)$$

$$\propto -\left(\frac{5-3n}{3-n}\right)\Delta M.$$ (5.77)

$\Delta M > 0$ (we are adding mass), so $\Delta\Omega < 0$ (i.e. star spins down) if $\frac{5-3n}{3-n} > 0$ (e.g. $n = \frac{3}{2}$), and $\Delta\Omega > 0$ (star spins up) if $\frac{5-3n}{3-n} < 0$ (e.g. $n = 2$). The latter case has a 'squashier' equation of state, so the star shrinks more and has to spin up to conserve angular momentum.

(ii) Consider a star in a binary system which loses mass onto its companion. The picture here is:

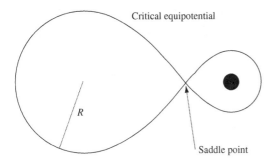

Critical equipotential

R

Saddle point

Fig. 5.4.

The curve shows an effective potential surface which includes the gravity from each star plus the centrifugal potential (in the frame of the stars). The critical equipotential is the one where either star will lose mass to the other if it expands beyond that surface, and the mass transfer takes place initially through the saddle point in the potential surface. In practice binary systems have to be rather close (periods of a few hours) for this to occur. In fact, since a binary star could never form if one of the stars was bleeding mass through the saddle point in the potential onto its companion, such behaviour (termed 'Roche lobe overflow') usually happens late in the lifetime of the binary when either the binary has become tighter through some mechanism and/or one of the components has swelled up when it turned into a red giant.

When the star loses mass ΔM by overflow then the initial adjustment follows the scaling relation at constant K, $R \propto M^{\frac{1-n}{3-n}}$. (Note we have neglected the deviation from spherical symmetry in the outer regions of the overflowing star.) Thus R increases for $3 > n > 1$. At the same time R_{crit} decreases (since there is less mass in the star), so the process is unstable and the mass transfer proceeds further. For the stars to be stable against mass transfer, R has to shrink more than R_{crit}, and this is not possible for $3 > n > 1$. For further details on mass exchange in binary stars see e.g. Hilditch, *An Introduction to Close Binary Stars* (Cambridge University Press, 2001).

5.9 Summary: general method for scaling relations

In order to find out how some property X scales with some property Y for polytropic stars of a given K, n, then:

- write down $X(\rho, r)$,

- convert to $X(\theta, \xi)$ using $\rho = \rho_c \theta^n$ and $\xi = \left(\dfrac{4\pi G \rho_c^{1 - \frac{1}{n}}}{K(n+1)} \right)^{\frac{1}{2}} r$, and

  hence find the dependence of X on ρ_c (which will generally have constants and an integral containing θ and ξ which are the same for all stars of this type),
- do likewise for Y,
- eliminate $\rho_c \longrightarrow X \leftrightarrow Y$.

Chapter 6
Propagation of sound waves

Sound waves play a central role in astrophysics, since they provide the principal mechanism by which disturbances propagate in fluids. (An exception to this is in the case of magnetised media where instead such information may be conveyed by Alfvén waves: see Chapter 13.) Astronomers routinely use the notion of a sound-crossing timescale in order to assess, for example, whether a given region has time to respond dynamically to a disturbance before it is affected in some other way (e.g. by heat diffusing into the region). This was precisely the sort of argument used in the last chapter where the relatively short time required for a sound wave to cross a star was used to deduce how a star responds to mass loss.

For the purpose of such order of magnitude estimates, it is often enough to use the speed of sound, familiar from elementary terrestrial physics, based on disturbances propagating in a uniform medium in which gravity is neglected. We start the chapter by reviewing the derivation of the wave equation in this simple case and in so doing, set up the apparatus that will be required in order to solve more complex examples.

6.1 Sound waves in a uniform medium

We start with the continuity and momentum equations in Eulerian form:

$$\frac{\partial \rho}{\partial t} + \boldsymbol{\nabla} \cdot (\rho \mathbf{u}) = 0, \tag{6.1}$$

$$\frac{\partial \mathbf{u}}{\partial t} + \mathbf{u} \cdot \boldsymbol{\nabla} \mathbf{u} = -\frac{1}{\rho} \boldsymbol{\nabla} p. \tag{6.2}$$

In the absence of external forces, the unperturbed state of the fluid, which is in equilibrium (i.e. $\frac{\partial}{\partial t} = 0$), is one of uniform density ρ_0, pressure p_0, and zero velocity ($\mathbf{u} = 0$). We then consider perturbations to this equilibrium, i.e.

$$p = p_0 + \Delta p, \tag{6.3}$$

$$\rho = \rho_0 + \Delta \rho, \tag{6.4}$$

$$\mathbf{u} = \Delta \mathbf{u}. \tag{6.5}$$

Before substituting these perturbed expressions into the fluid equations, we need to be careful because these perturbations are Lagrangian (i.e. they apply to individual fluid elements) whereas it is Eulerian perturbations (those which apply at a given location) which we must substitute into the Eulerian fluid equations (6.1) and (6.2). Thus if we consider some property of the flow (call it X) then the perturbation of X according to Equations (6.3)–(6.5) causes its value at a point P to change from its unperturbed value (X_0) for two reasons: (a) because the perturbation may have changed the value of X of the local fluid element and (b) because the perturbation may have moved a fluid element with a different *unperturbed* value of X so as to be located at point P. If the (small) *displacement* of a fluid element at P is denoted by the vector $\boldsymbol{\xi}$ then, to first order in small quantities, the change in X at point P is given by the sum of these contributions, i.e.

$$\delta X = \Delta X - \boldsymbol{\xi} \cdot \nabla X. \tag{6.6}$$

Note that we have here used the widespread convention of using Δ to represent Lagrangian perturbations and δ to represent Eulerian perturbations, i.e. those that must be substituted into the fluid equations.[1]

In the present case, the unperturbed quantities are all uniform, hence the second term on the right hand side of Equation (6.6) is zero and hence all δ quantities are equal to the Δ quantities. Since this is often the case in simple systems, elementary textbooks often gloss over this distinction when deriving the wave equation for the propagation of sound. However, we shall meet systems where the conversion contained in (6.6) is vital, and hence we introduce the general expression at this point.

---

[1] There is another convention in widespread use, whereby δ quantities (e.g. δX) are Lagrangian perturbations and primed quantities (e.g. X') are Eulerian perturbations. It is obviously essential to be clear on which convention is in use.

Substituting in the fluid equations, and retaining *only first order terms in the perturbed quantities* (i.e. those prefixed by Δ), we have:

$$\frac{\partial \Delta \rho}{\partial t} + \rho_0 \nabla \cdot (\Delta \mathbf{u}) = 0, \tag{6.7}$$

$$\frac{\partial \Delta \mathbf{u}}{\partial t} = -\frac{1}{\rho_0} \nabla \Delta p$$

$$= -\frac{\mathrm{d}p}{\mathrm{d}\rho} \frac{\nabla \Delta \rho}{\rho_0}. \tag{6.8}$$

Note that we have assumed in the latter equation that the fluid is barotropic and hence that a given change in density maps onto a unique change in pressure (and vice versa). In order to eliminate, say, $\Delta \mathbf{u}$ from these equations, and to obtain an equation for $\Delta \rho$, we $\frac{\partial}{\partial t}$ (6.7) $-\rho \nabla \cdot$(6.8) in order to obtain

$$\frac{\partial^2 \Delta \rho}{\partial t^2} = \frac{\mathrm{d}p}{\mathrm{d}\rho} \nabla^2 \Delta \rho. \tag{6.9}$$

This is a wave equation, i.e. in one dimension it has the solution

$$\Delta \rho = \Delta \rho_0 e^{i(kx - \omega t)}, \tag{6.10}$$

where ω is the angular frequency ($= 2\pi \nu$, where ν is the frequency in Hz) and k is the wavenumber ($= 2\pi/\lambda$ where λ is the wavelength). In higher dimensions

$$\Delta \rho = \Delta \rho_0 e^{i(\mathbf{k} \cdot \mathbf{x} - \omega t)}. \tag{6.11}$$

Substituting this solution into Equation (6.9) we see that $\frac{\omega^2}{k^2} = \frac{\mathrm{d}p}{\mathrm{d}\rho}$. Since $\frac{\omega}{k}$ is the speed of propagation of points of constant phase, the wave travels at a speed

$$c_\mathrm{s} = \sqrt{\frac{\mathrm{d}p}{\mathrm{d}\rho}}. \tag{6.12}$$

Similar equations can be readily derived for the other perturbed quantities. Since each quantity obeys the same wave equation, each can be written in the form of (6.11). By substituting such expressions back into (6.7) and (6.8) we can derive relationships between the perturbed quantities. In particular (in one dimension) the velocity

$$\Delta u = \frac{\Delta \rho_0}{\rho_0} \frac{\omega}{k} e^{i(kx - \omega t)} = \left(\frac{\Delta \rho}{\rho_0}\right) \frac{\omega}{k}. \tag{6.13}$$

We learn from this that the fluid velocity and density perturbations are in phase (because the ratio of the two is a real number) and we can readily envisage that this is the case by considering an instantaneous density maximum. Fluid at this location is surrounded on each side by elements whose displacement is of opposite sign, and hence the element at the centre of the density maximum has zero displacement from its equilibrium position. This element is thus passing through the origin of its oscillation, with maximum velocity.

We also learn from this that the amplitude of the velocity variation is $\frac{\Delta \rho_0}{\rho_0}$ times the propagation speed of the wave, c_s. However, since we only retained first order terms in our derivation of (6.7) and (6.8) we are only considering the case of small amplitude (linear) perturbations and it thus follows that $\Delta u_0 \ll c_s$. In other words, the disturbance propagates at a speed that far exceeds the speeds of individual fluid elements. (The same of course applies to all other examples of simple linear wave propagation, e.g. the speed with which a bowed violin string vibrates is much less than the propagation speed of the wave along the string.)

Sound waves propagate due to an interplay between density and velocity variations effected through the pressure term. We may immediately see why a sound wave *must* be a longitudinal wave since the stress tensor associated with pressure is diagonal, i.e. the forces acting on any surface in the fluid are perpendicular to that surface: evidently a transverse wave requires some non-diagonal terms in the stress tensor. (In this book, we defer consideration of viscous fluids until Chapter 11 but note here that a sufficiently viscous substance, such as treacle, can in principle support transverse waves too.) In detail, longitudinal waves propagate in fluids because a density perturbation gives rise to a pressure gradient, which then causes accelerations of the fluid elements. The resulting fluid velocities then induce further density perturbations and the disturbance propagates. Viewed this way, it is natural that the sound speed depends on the way that pressure forces react to changes in density, i.e. $\frac{dp}{d\rho}$. Thus a 'stiff' equation of state (high $\frac{dp}{d\rho}$) implies a large restoring force for small density perturbations and hence implies rapid propagation.

So far, we have not said how to calculate $\frac{dp}{d\rho}$. This needs to be done with the equation of state in mind, and we have seen in Chapter 4 how that depends on energy gains and losses in general. There are two obvious possibilities at either end of the energy transfer rate range. In the isothermal case $c_s^2 = \frac{dp}{d\rho}|_T$. Physically, isothermal sound waves occur if there is time over the timescale ω^{-1} (on which each fluid element executes its oscillation) for the rarefactions and compressions to pass heat to each other and so maintain a constant temperature. For an ideal gas, the isothermal sound speed is then

$$c_s = \left(\frac{\mathcal{R}_*}{\mu} T\right)^{\frac{1}{2}}.$$ (6.14)

In the adiabatic case $c_s^2 = \frac{dp}{d\rho}\big|_S$. At constant entropy there is no heat exchange between elements ($dQ = T\,dS$), so the heat transfer timescale is long compared with ω^{-1}. The compressions heat up and the rarefactions cool from $p\,dV$ work. Since we know that $p = K\rho^\gamma$ under these circumstances, we have

$$\frac{dp}{d\rho} = \gamma K \rho^{\gamma-1} = \gamma \frac{p}{\rho},$$ (6.15)

so

$$c_s = \left(\frac{\gamma \mathcal{R}_*}{\mu} T\right)^{\frac{1}{2}}.$$ (6.16)

Sound waves behave adiabatically or isothermally depending on the efficiency of heat leakage (by conduction or radiation). Which regime applies is obviously a matter of the physical conditions. For sound waves in air it turns out that $c_s = \left(\frac{\gamma \mathcal{R}_*}{\mu} T\right)^{\frac{1}{2}}$ (with γ appropriate for diatomic molecules) is a much better fit to experimental data than $c_s = \left(\frac{\mathcal{R}_*}{\mu} T\right)^{\frac{1}{2}}$, so sound waves in air are approximately adiabatic. Note that since the two sound speeds only differ by a factor of $\sqrt{\gamma}$, it often does not matter (for the purpose of order of magnitude calculations) which regime is appropriate.

A further point to note is that the thermal behaviour of the perturbations does not have to be the same as that of the unperturbed structure, since the timescales of interest can be very different. For example, as in the case of the Earth's atmosphere, the background atmosphere is roughly isothermal whereas, as noted above, sound waves propagate approximately adiabatically.

Before leaving this most simple example, we also point out that in this case c_s is not a function of the frequency ω. This means that all frequencies propagate at the same rate. Since any arbitrary disturbance can be represented as a Fourier sum of sinusoidal disturbances of various frequencies, this implies that in such a wave the *shape* of a packet of waves of different frequencies is therefore preserved as it propagates. For this reason waves for which c_s is not a function of ω are called *non-dispersive*.

6.2 Propagation of sound waves in a stratified atmosphere

What happens if there *are* external forces present – and therefore if the unperturbed structure is non-uniform? We examine this by consideration of sound waves propagating in an isothermal atmosphere with constant gravity, g, acting in the $-z$ direction. Evidently, the x and y components of the equation of motion are unaffected, so horizontal sound waves are unaffected by gravity. Thus considering the z-dependent terms only, we have

$$\frac{\partial \rho}{\partial t} + \frac{\partial}{\partial z}(\rho u) = 0,$$
$$\frac{\partial u}{\partial t} + u\frac{\partial u}{\partial z} = -\frac{1}{\rho}\frac{\partial p}{\partial z} - g \tag{6.17}$$

and the equilibrium is

$$u_0 = 0,$$
$$\rho_0(z) = \tilde{\rho}e^{-\frac{z}{H}}, \tag{6.18}$$

where the scale height $H = \frac{R_* T}{g\mu}$. Here we have used the subscript 0 to denote the equilibrium values, which are now functions of height in the atmosphere. Note also that $p_0(z) = \tilde{p}e^{-\frac{z}{H}}$.

We now perturb the medium, so $u \to \Delta u$, $\rho_0 \to \rho_0 + \Delta\rho$, and $p_0 \to p_0 + \Delta p$. We will be substituting these into (6.17), noting the general relationship between Eulerian and Lagrangian perturbations (6.6). We will also need two useful relationships between the perturbed quantities $\Delta\rho$ and $\Delta\mathbf{u}$ and the derivatives of the Lagrangian displacement, $\boldsymbol{\xi}$. The first,

$$\Delta\mathbf{u} = \frac{d\boldsymbol{\xi}}{dt} = \frac{\partial\boldsymbol{\xi}}{\partial t} + \mathbf{u}\cdot\nabla\boldsymbol{\xi}, \tag{6.19}$$

is obvious, i.e. the time derivative of the perturbed displacement of a particular fluid element is just the perturbed velocity of the element. Since in the present case the unperturbed velocity is zero, this just reduces to

$$\Delta\mathbf{u} = \frac{\partial\boldsymbol{\xi}}{\partial t}. \tag{6.20}$$

The second can be derived from the continuity equation in Lagrangian form:

$$\frac{d\rho}{dt} + \rho\nabla\cdot\mathbf{u} = 0. \tag{6.21}$$

Over a time Δt, the resulting change in density of a particular element can therefore be related to $\boldsymbol{\xi}$ by

$$\Delta\rho + \rho_0 \left(\nabla \cdot \frac{\partial \boldsymbol{\xi}}{\partial t} \right) \Delta t = 0, \tag{6.22}$$

and so we have

$$\Delta\rho + \rho_0 \nabla \cdot \boldsymbol{\xi} = 0. \tag{6.23}$$

We now substitute expressions for the Eulerian perturbations (i.e. $\delta \mathbf{u} = \Delta \mathbf{u}$, and since we are dealing with z-variations we can write $\delta\rho = \Delta\rho - \xi_z \frac{\partial \rho_0}{\partial z}$ and $\delta p = \Delta p - \xi_z \frac{\partial p_0}{\partial z}$, together with Equations (6.19) and (6.23)) into (6.17). The continuity equation becomes

$$\frac{\partial \Delta\rho}{\partial t} - \Delta u_z \frac{\partial \rho_0}{\partial z} + \Delta u_z \frac{\partial \rho_0}{\partial z} + \rho_0 \frac{\partial \Delta u_z}{\partial z} = 0, \tag{6.24}$$

i.e.

$$\frac{\partial \Delta\rho}{\partial t} + \rho_0 \frac{\partial \Delta u_z}{\partial z} = 0. \tag{6.25}$$

Likewise the momentum equation becomes

$$\frac{\partial \Delta u_z}{\partial t} = -\frac{1}{(\rho_0 + \Delta\rho - \xi_z \frac{\partial \rho_0}{\partial z})} \frac{\partial}{\partial z} \left(p_0 + \Delta p - \xi_z \frac{\partial p_0}{\partial z} \right) - g. \tag{6.26}$$

As usual, the unperturbed quantities cancel and we are left (to first order in perturbed quantities) with

$$\frac{\partial \Delta u_z}{\partial t} = \frac{1}{\rho_0^2} \left(\Delta\rho - \xi_z \frac{\partial \rho_0}{\partial z} \right) \frac{\partial p_0}{\partial z} - \frac{1}{\rho_0} \frac{\partial \Delta p}{\partial z} + \frac{\xi_z}{\rho_0} \frac{\partial^2 p_0}{\partial z^2} - \frac{\Delta\rho}{\rho_0^2} \frac{\partial p_0}{\partial z}. \tag{6.27}$$

The first and fifth terms cancel, as do the second and fourth (given the exponential form of the unperturbed pressure and density profiles). Thus we are simply left with

$$\frac{\partial \Delta u_z}{\partial t} = -\frac{1}{\rho_0} \frac{\partial \Delta p}{\partial z} = -\frac{c_u^2}{\rho_0} \frac{\partial \Delta\rho}{\partial z}, \tag{6.28}$$

where the latter equality implies that the perturbations obey a barotropic equation of state and where $c_u = \sqrt{\frac{dp}{d\rho}}$ is the usual sound speed in the case of a uniform medium. Note that if the unperturbed atmosphere is isothermal, and if the perturbations are either isothermal or adiabatic, then c_u is independent of z. We can see that (6.28) is correct by noting that for a given Lagrangian fluid element, the mass per unit area ($\rho \, dz$)

is conserved when the element is perturbed. Therefore the acceleration of that element changes only as a result of changes in the pressure difference across the element.

We now differentiate (6.25) with respect to time:

$$\frac{\partial^2 \Delta \rho}{\partial t^2} + \rho_0 \frac{\partial}{\partial z}\left(\frac{\partial \Delta u_z}{\partial t}\right) = 0, \tag{6.29}$$

which, from (6.28), becomes

$$\frac{\partial^2 \Delta \rho}{\partial t^2} - c_u^2 \frac{\partial^2 \Delta \rho}{\partial z^2} + \frac{c_u^2}{\rho_0}\frac{\partial \rho_0}{\partial z}\frac{\partial \Delta \rho}{\partial z} = 0. \tag{6.30}$$

Substituting from (6.18) we then finally obtain

$$\frac{\partial^2 \Delta \rho}{\partial t^2} - c_u^2 \frac{\partial^2 \Delta \rho}{\partial z^2} - \frac{c_u^2}{H}\frac{\partial \Delta \rho}{\partial z} = 0. \tag{6.31}$$

This becomes the usual wave equation as $H \to \infty$, as of course it must.
We now write

$$\Delta \rho \propto e^{i(kz - \omega t)} \tag{6.32}$$

and then find, if the differential equation is to be satisfied,

$$-\omega^2 = -c_u^2 k^2 + c_u^2 \frac{ik}{H}, \tag{6.33}$$

i.e.

$$\omega^2 = c_u^2\left(k^2 - \frac{ik}{H}\right). \tag{6.34}$$

This relation between the (angular) frequency ω and the wavenumber k is called the *dispersion relation*. It is a quadratic which we solve for $k(\omega)$.

$$k^2 - \frac{ik}{H} - \frac{\omega^2}{c_u^2} = 0 \tag{6.35}$$

$$\Rightarrow k = \frac{i}{2H} \pm \sqrt{\left(\frac{\omega}{c_u}\right)^2 - \left(\frac{1}{2H}\right)^2}. \tag{6.36}$$

Now if $\omega > \frac{c_u}{2H}$, then

$$\text{Im}(k) = \frac{1}{2H}, \tag{6.37}$$

$$\text{Re}(k) = \pm\sqrt{\left(\frac{\omega}{c_u}\right)^2 - \left(\frac{1}{2H}\right)^2},$$

and so

$$\Delta\rho \propto e^{-\frac{z}{2H}} e^{i\left(\pm\sqrt{\left(\frac{\omega}{c_u}\right)^2 - \left(\frac{1}{2H}\right)^2}\, z - \omega t\right)}. \tag{6.38}$$

If we set $\text{Re}(k) = K$, then the oscillatory part looks like the usual wave solution, where in this case we have

$$K^2 = \left(\frac{\omega}{c_u}\right)^2 - \left(\frac{1}{2H}\right)^2. \tag{6.39}$$

This implies that lines of constant phase propagate at $v_{\text{phase}} = \omega/K$, i.e.

$$v_{\text{phase}} = c_u \left[1 + \left(\frac{1}{2KH}\right)^2\right]^{\frac{1}{2}}. \tag{6.40}$$

Since this expression is a function of K and ω, the wave propagates dispersively (i.e. a wave packet composed of different frequencies changes shape as it propagates). We can explore this a bit further by considering two waves with frequencies $\tilde{\omega} \pm \delta\omega$ and wavenumbers $\tilde{k} \pm \delta k$. The superposed amplitude is then

$$\sin\left(\tilde{k}z - \tilde{\omega}t + (\delta k\, z - \delta\omega\, t)\right) + \sin\left(\tilde{k}z - \tilde{\omega}t - (\delta k\, z - \delta\omega\, t)\right)$$
$$= 2\sin\left(\tilde{k}z - \tilde{\omega}t\right)\cos(\delta k\, z - \delta\omega\, t). \tag{6.41}$$

The first factor is the usual wave propagating at the phase velocity, $\frac{\tilde{\omega}}{\tilde{k}}$, whereas the second represents the modulation of the wave amplitude by a slowly varying envelope function that propagates at speed $\frac{\delta\omega}{\delta k}$. In the limit, this becomes $\frac{d\omega}{dk}$, which is known as the group velocity, since it describes the velocity of the centroid of the wave packet (rather than the velocities of the individual waves that comprise the packet). In this case we have $v_{\text{group}} = \frac{d\omega}{dK}$; from (6.39) we see that

$$v_{\text{group}} v_{\text{phase}} = c_u^2 \tag{6.42}$$

and hence

$$v_{\text{group}} = c_u \left[1 + \left(\frac{1}{2KH}\right)^2\right]^{-\frac{1}{2}}. \tag{6.43}$$

We now turn to considering the imaginary part of k, which gives rise to the exponentially decaying amplitude with height in (6.38). We can obtain an expression for the corresponding velocity perturbation by substituting expressions of the form (6.32) into Equations (6.19) and (6.23):

$$\Delta \mathbf{u} = \frac{\Delta \rho}{\rho_0} \frac{\omega}{k}. \tag{6.44}$$

Since the amplitude of $\Delta \rho$ scales as $e^{-z/2H}$, whereas the background density scales as $e^{-z/H}$, we see that the perturbed velocity (and the fractional density variation) then *increases* with height according to $e^{+z/2H}$.

We can understand the growth of the velocity perturbations (and fractional density variations) with height as follows. If there is no energy dissipation (as must be the case, since our equations omit the physical mechanism – viscosity – by which energy is dissipated in fluids) then the kinetic energy flux in a wave is independent of the height. Thus $\rho_0 (\Delta u)^2 \frac{\omega}{k}$ is a constant. Thus if $\rho_0 \propto e^{-\frac{z}{H}}$ then the wave amplitude must grow in order to conserve kinetic energy flux in an atmosphere that is increasingly rarefied as the wave propagates upwards. There is nothing in our equations to apparently stop the fluid velocities rising without limit, until we recall the fact that once these velocities approach the sound speed, the perturbed density becomes comparable with the unperturbed density. At this point, our linear treatment breaks down: the non-linear wave would then steepen and form a shock (see Chapter 7).

Apparently, then, when you clap your hands, the upwardly propagating sound wave eventually generates a shock in the upper atmosphere! This would seem unlikely and is indeed prevented by the fact that viscosity in the atmosphere actually damps the wave before it becomes non-linear.

The whole of the above discussion was based on the assumption that $\omega > \frac{c_u}{2H}$. If, instead, $\omega < \frac{c_u}{2H}$ then k is wholly imaginary and > 0, so

$$\Delta u = e^{kz} e^{iwt}. \tag{6.45}$$

This has no term which allows propagation – it is a standing wave (i.e. all points oscillate in phase with an amplitude that varies with z). One is used to thinking of standing waves as resulting from superpositions of oppositely directed travelling waves (e.g. reflections off the ends of organ pipes). In this case, you might think of the standing wave as arising from reflections resulting from the changing properties of the atmosphere (note that as ω approaches $\frac{c_u}{2H}$ from above, the wavelength of the travelling wave becomes of the same order as H so that evidently changes in properties of the atmosphere over one wavelength start to become very significant).

Before leaving this example, we note one more thing. We introduced a certain amount of complexity into the problem by specifying Lagrangian perturbations, converting into the corresponding Eulerian equations, and then substituting into the Eulerian fluid equations. Why did we not simply start by casting the problem in terms of Eulerian perturbations? The reason is that the relationship between pressure and density perturbations is well defined for Lagrangian perturbations. For example, in the case of adiabatic perturbations each fluid element conserves its entropy and hence $\Delta p = \Delta \rho \times \gamma p_0 / \rho_0$. Since in an isothermal atmosphere the fluid elements at different heights have different entropies, when fluid elements are perturbed the entropy per unit mass *at a given location* is *not* preserved. Therefore $\delta p \neq \delta \rho \times \gamma p_0 / \rho_0$. In order to take advantage of a simple relationship between perturbed pressure and density, we therefore set up the problem in terms of Lagrangian perturbations.

6.3 General approach to wave propagation problems

The above has been an example of how to handle waves in general, and clearly the techniques can be applied in other circumstances. Here we summarise the general methodology.

Generally, for all wave equations
- write down the fluid equations,
- describe the equilibrium to be perturbed,
- choose the perturbed variables Δwhatever (where Δ is a Lagrangian perturbation, i.e. it refers to individual fluid elements),
- transform the Δquantities into δquantities (i.e. Eulerian perturbations referring to conditions at fixed location), using (6.6). If the unperturbed structure is not uniform, this will involve extra terms in ξ and Equations (6.19) and (6.22) (relating derivatives of ξ to the perturbed velocity and density) will turn out to be useful,
- substitute the δquantities (in terms of the Δquantities) into the Eulerian fluid equations and retain only first order terms in Δanything, (Note that unperturbed terms should cancel!)
- eliminate the Δvariables between equations so that the resultant equation is in terms of one variable only – this usually involves differentiating the equations you had,
- substitute in a solution of the form $\Delta \propto e^{i(kx - \omega t)}$,
- obtain a relation between ω and k – the dispersion relation.

In the case of a disturbance of real k, the dispersion relation indicates whether ω is real or imaginary and hence determines the stability of the system:

$\text{Re}(\omega) \rightarrow$ oscillating solution.
Negative $\text{Im}(\omega) \rightarrow$ exponentially decaying (damped) solution.
Positive $\text{Im}(\omega) \rightarrow$ exponentially growing (unstable) solution.

For real ω the dispersion relation indicates the spatial properties of the wave:

$\text{Re}(k) \rightarrow$ oscillating solution.
Positive $\text{Im}(k) \rightarrow$ exponentially decaying solution.
Negative $\text{Im}(k) \rightarrow$ exponentially growing solution.

The relationship between $\text{Re}(k)$ and ω gives the phase and group velocities, as above. (As a reminder: for a non-dispersive medium the phase and group velocities are the same.)

6.4 Transmission of sound waves at interfaces

Suppose we have a non-dispersive medium with a boundary to another at $x = 0$, and consider that the sound speed at $x < 0$ is c_{s1} and for $x > 0$ it is c_{s2}. We consider a sound wave in the x direction coming from $x < 0$ and ask what happens when it reaches and crosses the interface.

We suppose that the incident wave has a frequency ω. At the boundary the variables must be single valued and the accelerations finite, so the oscillations induced in the second medium must have the same frequency as in the first, and the wave amplitudes and their

Fig. 6.1.

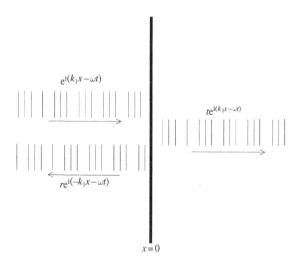

derivatives must be continuous across the interface. Given the frequency, the difference in sound speed is accounted for by having different ks for the sound waves in the two media. Therefore (setting the amplitude of the perturbed velocity in the incident wave equal to unity and the reflected and transmitted wave amplitude equal to r and t), continuity of wave amplitude and its derivative across the boundary implies

$$1 + r = t,$$
$$k_1(1 - r) = k_2 t,$$

(6.46)

$\Rightarrow$

$$t = \frac{2k_1}{k_1 + k_2},$$
$$r = \frac{k_1 - k_2}{k_1 + k_2}.$$

(6.47)

Note that the kinetic energy flux in a wave is $\propto \rho(\delta u)^2 c_s \sim \frac{p}{c_s}(\delta u)^2$, and p is constant across the interface. Hence (recalling that $c_s = \frac{\omega}{k}$), the initial energy is k_1 times a constant, and the energies taken by the reflected and transmitted waves are respectively the same constant times $k_1 \left(\frac{k_1 - k_2}{k_1 + k_2}\right)^2$ and $k_2 \left(\frac{2k_1}{k_1 + k_2}\right)^2$. The sum of these last two is k_1, so, as we would hope and expect, the energy carried in by the incident wave is the same as that carried away by the transmitted and reflected waves.

The above derivation illustrates some very general properties of waves propagating between different media. Any discontinuous change in sound speed produces a reflection at the boundary: if the sound speed increases at the boundary ($k_2 < k_1$) the reflected wave is in phase with the incident wave, whereas in the opposite case the wave undergoes a π phase shift at the boundary. In the extreme case of a large change in sound speed at the boundary, the reflected amplitude tends to ± 1 times the incident amplitude. For the case of a large decrease in sound speed at the boundary, the amplitude of the transmitted wave goes to zero.

In principle, then, one can see the possibility that sound waves may be almost totally reflected when crossing an interface between a hot and a much colder medium. This effect is important when one considers the propagation of sound waves between different phases of the interstellar medium (ISM) inasmuch as sound waves tend to bounce off cold, dense structures rather than propagating through them. Therefore it is difficult to excite disturbances in cold dense clouds – the so-called Giant Molecular Clouds – through the action of sound waves propagating in the surrounding low density medium. This is something of an issue, since Giant Molecular Clouds are composed of gas which is in a state of significant internal motion and it is unclear what is the origin of these internal motions. The above argument goes against the notion that supernovae exploding in the low density medium can set up disturbances that propagate effectively within the cloud. (Actually, in the case of the interstellar medium, the primary agent of propagation of disturbances is via magnetised (Alfvén) waves (see Chapter 13), but in this case also the propagation speed drops considerably within the high density Giant Molecular Clouds and so the same considerations apply.)

Chapter 7
Supersonic flows

In the last chapter, sound waves were introduced as the way that low amplitude (i.e. *linear*) disturbances propagate in a fluid. But what if a piece of fluid is subject to a non-linear disturbance, e.g. compression by a large factor or acceleration to velocities that are large compared with the sound speed? The result of such a disturbance is the propagation of a *shock*. On Earth, shocks are produced, for example, by the rapid pushing of a piston into a cold gas, or by the passage of a supersonic aircraft.

Astrophysics abounds in shock phenomena because gravity is an effective way of accelerating gas to high velocities: for example, gas free-falling onto the surface of stars or gas orbiting in a spiral galaxy like the Milky Way travels at hundreds of kilometres per second; in clusters of galaxies, the free-fall speed of the gas may attain thousands of kilometres per second. Such speeds correspond to the speed of sound in gas that is respectively at $\sim 10^6$ K and $\sim 10^8$ K; for gas that is cooler than this, any relative motion between fluid elements at these sorts of speeds must result in a shock. Note that acceleration to supersonic velocities (in the frame of the galaxy, say) does not itself generate a shock, since in its own rest frame the gas is at rest! It is only the sudden deceleration when it meets other fluid in relative motion that produces shock phenomena, e.g. the hot 'accretion spots' as gas streams down onto newly formed stars or where interstellar clouds collide with each other. Another thing that should be remembered by students who are used to estimating when shocks occur in terrestrial environments is that the dependence of sound speed on relative molecular weight ($c_s \propto \mu^{-1/2}$) is important when one goes from an atmosphere that is mainly composed of nitrogen (on Earth) to one composed mainly of atomic or molecular hydrogen in astronomical environments. The lighter molecules result in much higher sound speeds at given temperature (for example, the sound speed in Giant Molecular Clouds (GMCs), which are mainly molecular hydrogen, is comparable with the sound speed on Earth when the GMCs are at temperatures of a mere $10\,\mathrm{K}$ or so).

In the next chapter, we treat one particular astronomical example where shocks are important (the propagation of a blast wave from a supernova through the interstellar medium) in some detail. Here we now set up the basic theory that is applicable to all shocks.

7.1 Shocks

Disturbances always propagate at a speed c_s relative to the fluid. Consider an observer situated at the source of a spherical disturbance watching the fluid flow past it at a speed v. The velocity of the disturbance *relative to the observer* is just the vector sum of v and the velocity vectors (of magnitude c_s) of the disturbance relative to the fluid. As may be seen from Figure 7.1, for *subsonic* flow the resultant vector always sweeps out 4π steradians as seen from the point of view of the observer.

For supersonic flow the resultant vector is always to the right. Figure 7.2 demonstrates that there is a maximum angle that the resultant makes with the flow direction of the fluid. One can consider the superposition of the vector v with disturbances whose angle to the direction of v varies from π (upstream propagation) to zero (downstream propagation). As this angle is initially reduced from π, the angle between the resultant and v first of all increases, but attains a maximum when the resultant is tangential to the sphere of radius c_s. From the diagram one sees that this maximum angle is α such that $\sin\alpha = c_s/v$. This

Fig. 7.1.

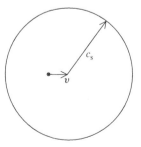

Fig. 7.2.

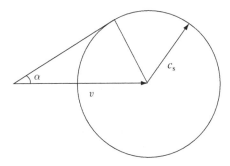

delineates a 'Mach cone' of directions in which disturbances from the point can propagate. We call the ratio of the flow speed to the sound speed the

$$\text{Mach number } M = \frac{v}{c_{\mathrm{s}}}.$$

For an obstacle in a supersonic flow, the disturbances cannot propagate upstream from the obstacle, i.e. the flow cannot adjust to the presence of the obstacle because there is no way of propagating a signal in that direction. Therefore the flow is undisturbed until it reaches the obstacle, and there its properties change discontinuously in a *shock*.

When the flow is subsonic, by contrast, it can adjust to the presence of an obstacle because the disturbances it causes can be communicated upstream through the fluid.

We need to consider the fluid flow through the shock, which manifests itself as a boundary between two regions of the fluid where the conditions change discontinuously. The usual conditions like the conservation of matter, momentum and energy equations have to apply here, just as in any fluid, so all we have to do is interpret them as they apply across the shock front. It is easiest to do this in the frame of the shock.

The equation of continuity (taking the normal to the shock (= surface of discontinuity) in the x direction) is

$$\frac{\partial \rho}{\partial t} + \frac{\partial}{\partial x}(\rho u_x) = 0. \tag{7.1}$$

We can integrate this over a layer of thickness $\mathrm{d}x$ around the shock, to obtain

$$\frac{\partial}{\partial t} \int \rho \, \mathrm{d}x + \rho u_x|_{\frac{\mathrm{d}x}{2}} - \rho u_x|_{-\frac{\mathrm{d}x}{2}} = 0. \tag{7.2}$$

Fig. 7.3.

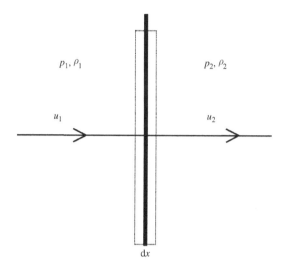

For an infinitesimal layer dx the mass flux in = mass flux out (i.e. mass does not accumulate in that layer), so $\frac{\partial}{\partial t}\int\rho\,dx = 0$ and then ρu_x must be the same on both sides of the shock. So

$$\rho_1 u_1 = \rho_2 u_2. \tag{7.3}$$

The momentum equation is dealt with similarly. We use the momentum equation in conservative form since that has differentials for everything (and so the integration is easier), and then we have for the x component

$$\frac{\partial}{\partial t}(\rho u_x) = -\frac{\partial}{\partial x}(\rho u_x u_x + p) - \rho\frac{\partial}{\partial x}\Psi. \tag{7.4}$$

Then integrating over the small region dx gives

$$\frac{\partial}{\partial t}\int(\rho u_x)\,dx = -(\rho u_x u_x + p)|_{\frac{dx}{2}} + (\rho u_x u_x + p)|_{-\frac{dx}{2}} \tag{7.5}$$

since the contribution from the gravitational term is negligible because Ψ is continuous at the shock. Consequently

$$\rho_1 u_1^2 + p_1 = \rho_2 u_2^2 + p_2, \tag{7.6}$$

i.e. the sum of the thermal and ram pressures across the boundary is a constant.

(The other two velocity components u_y and u_z do not change across the boundary. This comes about because $\rho u_x u_y$ and $\rho u_x u_z$ are constant across the boundary from the momentum equation, and continuity gives us ρu_x is constant across the boundary, and therefore u_y and u_z must also be the same on either side of the boundary. As a consequence we can, and do, take a coordinate frame in which the velocities tangential to the shock front are zero.)

The energy equation is dealt with similarly, though here we have to make some assumption about the nature of cooling in the shocked gas. We first assume that the gas cannot cool, i.e. that the shock is adiabatic (so $\dot{Q}_{cool} = 0$) and consider another limit (where the gas is isothermal) in Section 7.2 below. For an adiabatic shock, the result from the energy equation (4.32) is then that $(E + p)u$ is the same on both sides of the shock, and so across the shock front

$$\left[\frac{1}{2}u^2 + \Psi + \mathcal{E} + \frac{p}{\rho} \right] \rho u = \text{constant}. \tag{7.7}$$

We know from (7.3) that ρu is constant across the shock, and Ψ is a continuous function at the shock, so

$$\frac{1}{2}u_1^2 + \mathcal{E}_1 + \frac{p_1}{\rho_1} = \frac{1}{2}u_2^2 + \mathcal{E}_2 + \frac{p_2}{\rho_2}. \tag{7.8}$$

These three conditions are the Rankine–Hugoniot relations. The first is simply mass conservation, the second expresses the fact that a shock represents a conversion of ram pressure to thermal pressure, and the third the conversion of kinetic energy into enthalpy. Qualitatively, a shock converts an ordered flow upstream into a disordered (i.e. hot) flow downstream. (Note however that although the Rankine–Hugoniot conditions are reversible, i.e. would in principle allow the opposite process of a hot flow being transformed into a cold ordered flow, this outcome is actually prohibited by entropy considerations: see below.)

It is convenient to re-write the third Rankine–Hugoniot condition so as to re-cast the internal energy in terms of other thermodynamic quantities. This is done by noting that *if* any gas is brought to a particular point in thermodynamic phase space by adiabatic compression, then the internal energy stored in the gas is just the $p\,dV$ work done along that adiabatic path. Since an adiabatic path is defined by $p = K\rho^\gamma$ (for some constant K), the $p\,dV$ work is simply $K\rho^\gamma\,d(1/\rho)$ which can be written as $\frac{1}{(\gamma-1)}\frac{p}{\rho}$. However, the internal energy is defined as a function of thermodynamic variables, regardless of how the fluid arrives in that state. Hence, although we imagined the fluid arriving at given p and ρ along an adiabatic path in order to readily calculate the internal energy in that case, this expression for the internal energy per unit mass is universally valid, i.e.

$$\mathcal{E} = \frac{1}{(\gamma - 1)} \frac{p}{\rho}.$$ (7.9)

Assuming that γ does not change across the shock, Equation (7.8) can be re-written as

$$\frac{1}{2}u_1^2 + \frac{\gamma}{\gamma - 1}\frac{p_1}{\rho_1} = \frac{1}{2}u_2^2 + \frac{\gamma}{\gamma - 1}\frac{p_2}{\rho_2}.$$ (7.10)

Then, since the sound speed satisfies $c^2 = \gamma\frac{p}{\rho}$, we have

$$\frac{1}{2}u_1^2 + \frac{c_1^2}{\gamma - 1} = \frac{1}{2}u_2^2 + \frac{c_2^2}{\gamma - 1}.$$ (7.11)

The Rankine–Hugoniot relations thus derived are simple, elegant and symmetrical, but unfortunately we have to do some rather laborious algebra in order to derive from them some physically interesting properties of the flow. This algebraical manipulation is notoriously likely to end up in fruitless dead ends unless one charts a rather clear path. Although the path taken below is not the only one, it has the advantage of retaining the symmetry of the expressions as long as possible and is to be recommended.

Using Equation (7.3), set

$$j = \rho_1 u_1 = \rho_2 u_2,$$ (7.12)

and the other two relations then become

$$p_1 + \frac{j^2}{\rho_1} = p_2 + \frac{j^2}{\rho_2}$$ (7.13)

for Equation (7.6), and

$$\frac{1}{2}\frac{j^2}{\rho_1^2} + \frac{\gamma}{(\gamma - 1)}\frac{p_1}{\rho_1} = \frac{1}{2}\frac{j^2}{\rho_2^2} + \frac{\gamma}{(\gamma - 1)}\frac{p_2}{\rho_2}$$ (7.14)

for Equation (7.8).

This again assumes nothing dramatic has happened to the nature of the gas in the shock to change the value of γ (e.g. by disassociation of molecules), but is normally true. Then, from (7.13) we have

$$j^2 = \frac{p_2 - p_1}{\frac{1}{\rho_1} - \frac{1}{\rho_2}},$$ (7.15)

and we now substitute this expression for j^2 in (7.14). This gives

$$\frac{1}{2}\left(\frac{p_2-p_1}{\frac{1}{\rho_1}-\frac{1}{\rho_2}}\right)\left[\frac{1}{\rho_1^2}-\frac{1}{\rho_2^2}\right]=\frac{\gamma}{(\gamma-1)}\left[\frac{p_2}{\rho_2}-\frac{p_1}{\rho_1}\right] \tag{7.16}$$

$\Rightarrow$

$$\frac{1}{2}(p_2-p_1)\left[\frac{1}{\rho_1}+\frac{1}{\rho_2}\right]=\frac{\gamma}{(\gamma-1)}\left[\frac{p_2}{\rho_2}-\frac{p_1}{\rho_1}\right] \tag{7.17}$$

$\Rightarrow$

$$\frac{1}{\rho_2}\left[\frac{1}{2}p_2-\frac{1}{2}p_1-\frac{\gamma}{(\gamma-1)}p_2\right]=\frac{1}{\rho_1}\left[\frac{1}{2}p_1-\frac{1}{2}p_2-\frac{\gamma}{(\gamma-1)}p_1\right] \tag{7.18}$$

$\Rightarrow$

$$\frac{1}{\rho_2}\left[\left(\frac{\gamma+1}{\gamma-1}\right)p_2+p_1\right]=\frac{1}{\rho_1}\left[\left(\frac{\gamma+1}{\gamma-1}\right)p_1+p_2\right] \tag{7.19}$$

and so

$$\boxed{\frac{\rho_2}{\rho_1}=\frac{(\gamma+1)p_2+(\gamma-1)p_1}{(\gamma+1)p_1+(\gamma-1)p_2}\quad\left(=\frac{u_1}{u_2}\right).} \tag{7.20}$$

In this form the equations are quite useful, and we can draw some general conclusions. In the limit of *strong* shocks, $p_2 \gg p_1$ (i.e. we can neglect the upstream pressure), we have

$$\frac{\rho_2}{\rho_1}\to\frac{\gamma+1}{\gamma-1} \tag{7.21}$$

and so, for $\gamma=\frac{5}{3}$, the density ratio pre- to post-shock is then 4. For weaker shocks the density ratio is less than this (it is obviously unity when $p_2=p_1$), so this gives us a maximum possible density contrast across adiabatic shocks in a monatomic gas of a factor of 4. Qualitatively, this behaviour arises because as the shock strength is increased to higher Mach numbers ($M=\frac{u_1}{c_1}$) the thermal pressure behind the shock increases more, and therefore stops the post-shock gas from being compressed much.

Note that in general $\frac{p_1}{\rho_1^\gamma}\neq\frac{p_2}{\rho_2^\gamma}$. This is obvious for strong shocks, where $p_2 \gg p_1$ but ρ_2 can be at most a factor $4\times\rho_1$. This means that in a shock, the gas jumps from one adiabat to another one of higher entropy.

It may be a little puzzling to the reader that we have specified that the shock is adiabatic and yet we have ended up, through use of this assumption, with the result that the gas jumped from one adiabat to another (i.e. from one value of K to another). The reason why this is confusing is that one often talks interchangeably of adiabatic changes in the sense of $dQ_{cool} = 0$ and in the sense of an equation of state $p = K\rho^\gamma$ (in fact we showed that these statements were equivalent in Chapter 4) starting from the first law of thermodynamics (Equation (4.3)). However, we emphasised that this equality only holds for *reversible* changes (i.e those in which there is no conversion of kinetic energy into heat energy). In the presence of viscosity, however, a change that is adiabatic (in the sense that $dQ_{cool} = 0$) does *not* imply that the before and after states are simply linked by the relation $p = K\rho^\gamma$ with K constant (i.e does not imply that fluid elements preserve their entropy).

This entropy jump across the shock is what allows us to define an arrow of time for the problem. It permits shocks as described (cold and fast $\rightarrow$ hot and slow in the frame of the shock) but not the opposite causality, because that would imply a *reduction* in entropy.

The above discussion shows us that viscosity plays a central role in shocks and yet we have not included any viscous terms in either the momentum or the energy equation! This is actually not a problem because we have applied the equations in integral form to find the change in conditions between two regions which are each inviscid. The fact that there is a region sandwiched in between where viscosity is important does not undermine the validity of these jump conditions, but just implies that, whatever the detailed structure of the region in between, it must somehow change the flow so as to fulfil the jump conditions.

Therefore if we are interested in the before and after properties of shocks, we do not necessarily have to concern ourselves with the details of the thin dissipative (viscous) region immediately downstream of the shock front. An exception to this is if we are engaged in modelling shocks in numerical simulations. One approach is to simply fit the jump conditions (the Rankine–Hugoniot conditions) as boundary conditions on the flow variables at the shock front. Alternatively, one can let the shock be handled by the normal equations of motion; in this case, it is however imperative to add a so-called artificial viscosity to the code in order to achieve the necessary dissipation (entropy generation) that must occur in a shock. This is most easily seen in the case of particle-based codes where, in the absence of such viscosity, particles would simply pass through each other in the shock region because, since they approach each other supersonically, there is no opportunity for them to learn about other particles in advance through the action of pressure forces.

7.2 Isothermal shocks

In general, $\dot{Q}$ will not be zero so eventually the shocked gas will cool, in some cases back to close to its original temperature. So a temperature profile through a shock could look like:

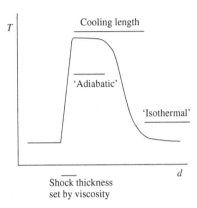

Fig. 7.4.

Provided the flow downstream is in a steady state the ρu and $p + \rho u^2$ are constant in the shocked flow. Therefore, the isothermal portion of the flow also obeys the first two Rankine–Hugoniot relations (7.3) and (7.6), despite the fact that there is an adiabatic portion sandwiched between it and the shock discontinuity. Another way of looking at it is to note that on the scale of interest in an isothermal shock, the discontinuity contains the adiabatic and cooling regions.

Adiabatic shocks are therefore those in which the cooling length ℓ_{cool} > the size of the system, and isothermal shocks are those for which ℓ_{cool} ≪ the lengthscale for the system.

In *isothermal* shocks the third Rankine–Hugoniot condition is no longer appropriate since heat is lost from the gas. Condition (7.8) is replaced by the simple condition that the flow has returned to its original temperature, i.e.

$$T_2 = T_1. \tag{7.22}$$

Since $T_1 = T_2$ we know that the isothermal sound speed c_s is the same both before and after the shock, and on each side $p = \rho c_s^2$. This relation and the Rankine–Hugoniot conditions,

$$\rho_1 u_1 = \rho_2 u_2, \tag{7.23}$$

$$\rho_1 u_1^2 + p_1 = \rho_2 u_2^2 + p_2, \tag{7.24}$$

now describe the relationship between the pre- and post-shock flow. Equation (7.24) $\Rightarrow$

$$\rho_1(u_1^2 + c_s^2) = \rho_2(u_2^2 + c_s^2), \tag{7.25}$$

which, with (7.23), $\Rightarrow$

$$(u_2 - u_1)c_s^2 = u_1 u_2(u_2 - u_1). \tag{7.26}$$

So, provided $u_2 \neq u_1$ (i.e. there is a shock), we have

$$c_s^2 = u_1 u_2. \tag{7.27}$$

Therefore

$$\frac{\rho_2}{\rho_1} = \frac{u_1}{u_2} = \left(\frac{u_1}{c_s}\right)^2 = M^2. \tag{7.28}$$

So in an isothermal shock, the compression factor in the shock is the square of the Mach number of the pre-shocked flow. So, unlike the adiabatic case where the compression factor was limited to a small number, the density ratio can have very high values in a high Mach number isothermal shock.

For an *isothermal* shock we have $c_s^2 = u_1 u_2$, so if $u_1 > c_s$ (i.e. if there is a shock at all) then $u_2 < c_s$. So the flow behind the shock is subsonic relative to the shock.

We can show that the post-shocked gas being subsonic is a general result for adiabatic shocks too. The derivation is another example of slightly tedious manipulation of the Rankine–Hugoniot conditions to produce a physically meaningful result. For a given γ (which does not change across the shock, by assumption) the Rankine–Hugoniot relations can be expressed in terms of the sound speeds $c^2 = \gamma p/\rho$ rather than the pressure p:

$$\rho_1 u_1 = \rho_2 u_2, \tag{7.29}$$

$$\rho_1 u_1^2 + \frac{\rho_1 c_1^2}{\gamma} = \rho_2 u_2^2 + \frac{\rho_2 c_2^2}{\gamma}, \tag{7.30}$$

$$\frac{1}{2}u_1^2 + \frac{c_1^2}{\gamma - 1} = \frac{1}{2}u_2^2 + \frac{c_2^2}{\gamma - 1}. \tag{7.31}$$

If you prefer to use Mach numbers, $u_1 = M_1 c_1$ and $u_2 = M_2 c_2$, so:

$$\rho_1 M_1 c_1 = \rho_2 M_2 c_2, \tag{7.32}$$

$$\rho_1 c_1^2 \left(M_1^2 + \frac{1}{\gamma} \right) = \rho_2 c_2^2 \left(M_2^2 + \frac{1}{\gamma} \right), \tag{7.33}$$

$$c_1^2 \left(M_1^2 + \frac{2}{\gamma - 1} \right) = c_2^2 \left(M_2^2 + \frac{2}{\gamma - 1} \right). \tag{7.34}$$

Dividing Equation (7.33) by (7.32) gives

$$M_2 c_1 \left(M_1^2 + \frac{1}{\gamma} \right) = M_1 c_2 \left(M_2^2 + \frac{1}{\gamma} \right), \tag{7.35}$$

which, with (7.34), gives

$$M_1^2 \left(M_2^2 + \frac{1}{\gamma} \right)^2 \left(M_1^2 + \frac{2}{\gamma - 1} \right) = M_2^2 \left(M_1^2 + \frac{1}{\gamma} \right)^2 \left(M_2^2 + \frac{2}{\gamma - 1} \right). \tag{7.36}$$

So we now have a general expression linking the pre-shock Mach number with the post-shock one. Expanding this gives some cancellations, which result in

$$\left[2M_1^2 M_2^2 \left(\frac{1}{\gamma - 1} - \frac{1}{\gamma} \right) - \frac{1}{\gamma^2} (M_1^2 + M_2^2) - \frac{2}{\gamma^2 (\gamma - 1)} \right] (M_1^2 - M_2^2) = 0. \tag{7.37}$$

So if $M_1^2 \neq M_2^2$, i.e. if there is a shock at all, then the term in square brackets must be zero. Thus we have, after simplifying a little further,

$$M_2^2 = \frac{2 + (\gamma - 1)M_1^2}{2\gamma M_1^2 - (\gamma - 1)}. \tag{7.38}$$

We can see that if $M_1 > 1$ then $M_2 < 1$ by setting $\gamma = 1 + \frac{1}{n}$, since then

$$M_2^2 = \frac{M_1^2 + 2n}{2M_1^2 - 1 + 2nM_1^2} \tag{7.39}$$

and $2M_1^2 - 1 > M_1^2$ and $2nM_1^2 > 2n$. So for any adiabatic shock with constant γ the post-shock flow is subsonic.

In fact, although we have only proved this for adiabatic shocks with constant γ and isothermal shocks, the fact that the post-shock flow is subsonic with respect to the shock is a general result. A little reflection reveals why this is so. The pre-shock flow is oblivious of the fact that it is approaching the shock, because it does so supersonically. The post-shock flow, instead, is the *result* of processing the flow in the shock front and it must therefore 'know' about conditions at the front in order to set conditions like the mass flux rate and momentum flux rate. It can only communicate with the shock front, however, if its motion is subsonic with respect to the shock front. In other words, in order to preserve causality in the shock, it is necessary for the shocked flow to emerge from the shock at subsonic speeds.

Chapter 8
Blast waves

One of the most important applications of shock wave theory in astronomy is to the problem of a supernova exploding in an interstellar cloud. Supernovae dump around 10^{44} Joules of thermal and kinetic energy of ejecta into a small region around the star on an astronomically minute timescale (a day or so). The shocked medium expands and sweeps up more gas, creating a large bubble in the interstellar medium (ISM). It is now known that the continuous injection of energy into the ISM in this way by successive supernovae is responsible for giving the ISM a 'Swiss cheese' structure, with bubbles of hot gas alternating with sheets and filaments of cooler material swept up in the cavity walls. Astronomical opinion is divided as to whether the net effect of supernova action is to suppress further star formation (through hot gas in the bubbles breaking out of the galactic potential) or to promote it (through enhanced fragmentation of gas swept up in the bubble walls). At any rate, the effect of supernovae is evidently fundamental to understanding the ISM in galaxies. Before we are in a position to estimate the global effects of successive supernovae, we must first construct an idealised model of a single supernova exploding in a uniform medium.

The title of the next section however indicates that the problem was first studied in another and more menacing context: the impetus for developing the elegant theory laid out below came from the need to model the effects of thermonuclear weapons exploding in the earth's upper atmosphere.

8.1 Strong explosions in uniform atmospheres

We consider an idealised explosion where an explosion energy E is delivered instantaneously at a point which is surrounded by an atmosphere of uniform density ρ_0, as illustrated in Figure 8.1.

Fig. 8.1.

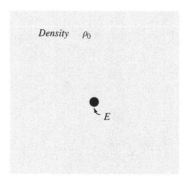

Density ρ_0

E

We will ignore the temperature of the atmosphere (so $T_0 = 0$ effectively) and so ignore any effects of thermal pressure from the medium confining the explosion. We can approach the problem in two ways – an approximate method where we can see what physics is going on, and a neat mathematical method which gives little insight. Naturally, we'll do both, starting with the physical approach, which will give the right sort of power law scalings even if it involves some apparently rather arbitrary assumptions that can only be justified *a posteriori*. We will then go on to present the full solution which will allow us to assess the validity of the approximations made.

8.1.1 Approximate method

As the explosion propagates out there will be a shock, and the medium is swept up into a shell of shocked gas with the explosion products behind it, as in Figure 8.2.

Fig. 8.2.

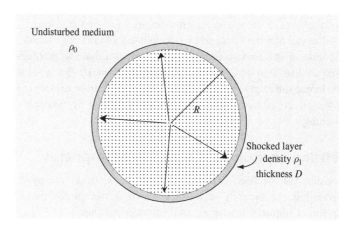

Undisturbed medium
ρ_0

R

Shocked layer
density ρ_1
thickness D

The temperature in the unshocked gas has been taken to be zero, so the Mach number for the shock $M \rightarrow \infty$. Thus for an adiabatic shock,

$$\frac{\rho_1}{\rho_0} = \frac{\gamma + 1}{\gamma - 1}. \tag{8.1}$$

We can calculate the thickness D of the layer of shocked gas on the assumption that all the mass in the gas is swept up into it, i.e.

$$\frac{4\pi}{3}\rho_0 R^3 = 4\pi \rho_1 R^2 D \tag{8.2}$$

if D is small enough. This implies

$$D = \frac{1}{3}\left(\frac{\gamma - 1}{\gamma + 1}\right) R. \tag{8.3}$$

For $\gamma = \frac{5}{3}$, $D/R \sim 0.08 \ll 1$, so the assumption of a thin shell is justified. [Any other realistic γ gives even thinner shells – e.g. diatomic molecules, $\gamma = \frac{7}{5}$ gives a density contrast of a factor 6, and so a thinner shell.]

Now we assume that all the gas in the shell moves with the same velocity at any instant. Effectively this is saying that because the shell is thin, we can use an average velocity to represent the gas motion within it. Then, in the frame of the shock the situation is as shown in Figure 8.3.

The Rankine-Hugoniot relation (7.3) (from the continuity equation) gives us

$$\rho_0 u_0 = \rho_1 u_1, \tag{8.4}$$

so for a strong shock

$$u_1 = \frac{\rho_0}{\rho_1} u_0 = \left(\frac{\gamma - 1}{\gamma + 1}\right) u_0. \tag{8.5}$$

Fig. 8.3.

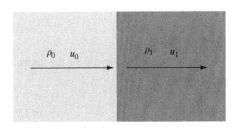

Relative to the unshocked gas (which is the same as being relative to the centre of the explosion), the velocity of the shocked gas U is

$$U = u_0 - u_1 = \frac{2u_0}{\gamma + 1} \tag{8.6}$$

If we now look at the radial momentum, we see that as the shell grows it gains momentum at a rate

$$\frac{d}{dt}\left[\frac{4\pi}{3}\rho_0 R^3 \frac{2u_0}{\gamma + 1}\right], \tag{8.7}$$

since the mass of gas swept up is $\frac{4\pi}{3}\rho_0 R^3$ and its velocity is $\frac{2u_0}{\gamma+1}$. This radial momentum gain has to be provided by the pressure acting on the inside of the shell, p_{in}. We have set the outside pressure to zero by setting $T_0 = 0$, so don't have to concern ourselves with any opposing pressure from the undisturbed medium.

Now we set

$$p_{\text{in}} = \alpha p_1 \tag{8.8}$$

where p_1 is the pressure in the shell i.e. we suppose the pressure acting on the inside of the shocked gas shell scales in some way with the pressure within the shell itself. This is the part of the derivation that appears pretty arbitrary, but we will make this assumption and examine the solution we get in this case.

For a strong shock we know

$$p_1 = \frac{2}{\gamma + 1}\rho_0 u_0^2, \tag{8.9}$$

so the rate of change of radial momentum equals the force over the whole area. Thus

$$\frac{d}{dt}\left[\frac{4\pi}{3}\rho_0 R^3 \frac{2u_0}{\gamma + 1}\right] = 4\pi\alpha R^2 \frac{2}{\gamma + 1}\rho_0 u_0^2 \tag{8.10}$$

There is a fair amount of cancellation here, so equation (8.10) becomes

$$\frac{d}{dt}\left[R^3 u_0\right] = 3\alpha R^2 u_0^2 \tag{8.11}$$

But u_0 is the speed with which the shock advances on the undisturbed gas, and so

$$u_0 = \frac{dR}{dt} \tag{8.12}$$

and hence

$$\frac{d}{dt}\left[R^3\dot{R}\right] = 3\alpha R^2\dot{R}^2. \tag{8.13}$$

We now make a further trial assumption, suggested by the homogeneous nature of the equations, that the solution is of the form $R \propto t^b$, and substitute. Then equation (8.13) becomes

$$b(4b-1)t^{4b-2} = 3\alpha b^2 t^{4b-2}, \tag{8.14}$$

We discount the solution $b = 0$ as being uninteresting, since then r is constant and see that otherwise

$$b = \frac{1}{4-3\alpha} \tag{8.15}$$

$\Rightarrow$

$$R \propto t^{\frac{1}{4-3\alpha}} \tag{8.16}$$

and

$$u_0 \propto t^{\frac{3\alpha-3}{4-3\alpha}} \propto R^{3\alpha-3}. \tag{8.17}$$

So under these circumstances we have power law scalings which tell us how the radius of the shell grows as a function of time in terms of some unknown quantity α.

In order to determine α we need to consider the energetics of the explosion. For an adiabatic blast wave, the energy of the explosion, E, is conserved, and is taken up in two ways:

• the kinetic energy of the shell, which is

$$\frac{1}{2}\frac{4\pi}{3}\rho_0 R^3 U^2, \tag{8.18}$$

• internal energy.

Now the internal energy per unit volume is $p/(\gamma-1)$ (see the expression for $\mathcal{E}$, the internal energy per unit mass, equation 7.9), and, since the shell is thin, most of the volume of the bubble created by the blast wave is in the internal cavity. With this in mind, we expect that most of the internal energy is in the material in the cavity (which contains little mass). Therefore the internal energy is approximately

$$\frac{4\pi}{3}R^3\frac{p_{\text{in}}}{\gamma-1} = \frac{4\pi}{3}R^3\frac{\alpha p_1}{\gamma-1}. \tag{8.19}$$

Therefore, using equation (8.18) for the kinetic energy term with equation (8.6) to substitute for U, and equation (8.9) for p_1, gives us

$$E = \frac{4\pi}{3} R^3 \left[\frac{1}{2} \rho_0 \left(\frac{2u_0}{\gamma+1} \right)^2 + \frac{\alpha}{\gamma-1} \frac{2\rho_0 u_0^2}{\gamma+1} \right], \qquad (8.20)$$

i.e.

$$E \propto R^3 u_0^2 \propto t^{\frac{6\alpha-3}{4-3\alpha}}. \qquad (8.21)$$

However, we know that energy is conserved, so we must have $6\alpha - 3 = 0$, i.e.

$$\alpha = \frac{1}{2} \qquad (8.22)$$

So, substituting back into (8.9) and (8.17) gives us

$$R \propto t^{\frac{2}{5}}, \qquad u_0 \propto t^{-\frac{3}{5}}, \qquad p_1 \propto t^{-\frac{6}{5}}. \qquad (8.23)$$

The main disadvantage of the above approach is that we don't know if it is correct in detail. We have drawn on the various physical properties, and shock conditions, to develop the form of the solution, but the main weak link is the α which linked the pressure inside the shocked gas to the pressure acting on it from the explosion products. They are the only two pressures in the whole model (since we set the outside pressure to zero), so there will be some link, but we have not justified assuming the ratio is constant, nor investigated what happens if it is not. [The real justification is that it gives the *right answer*! To know we have the right answer, we have to solve the equations more rigorously in some way.]

8.1.2 Similarity solution

This is short, neat and mathematical – and gives no insight into what is going on!

The problem as posed has only two parameters, E and ρ_0, and these cannot be combined to give quantities with dimensions of either length [L] or time [T]. If the problem is well-posed, this means there are no obvious characteristic lengthscales or timescales for the blast wave problem. To try to see what this means, let us suppose that λ is a scale parameter giving the size of the blast wave at a time t after the explosion. Apart from being a monotonically increasing function of time, the evolution of λ may depend on E and ρ_0. Now the only way of combining E, ρ_0 and t to get a dimension of length is

$$\lambda = \left(\frac{Et^2}{\rho_0}\right)^{\frac{1}{5}}.$$

(8.24)

What we can now do is introduce a dimensionless distance parameter in place of the radius r,

$$\xi = \frac{r}{\lambda} = r\left(\frac{\rho_0}{Et^2}\right)^{\frac{1}{5}},$$

(8.25)

and suppose, since there is no natural length-scale, that the solution is <u>self-similar</u>. What that means is that we assume that all variables in the problem, which we denote generically by $X(r, t)$, can be written in the form

$$X = X_1(t)\tilde{X}(\xi).$$

(8.26)

This implies is that if you take a snapshot of the distribution X at any time and plot it as a function of ξ rather than as a function of r, then it always has the same shape but is scaled up or down by a time-dependent factor $X_1(t)$. So, for example, the situation at two different times is as illustrated in Figure 8.4.

You find a similarity solution by substituting $X_1(t)\tilde{X}(\xi)$ for X into the relevant equations, remembering that $\xi = \xi(r, t)$, so

$$\frac{\partial X}{\partial r} = X_1 \frac{d\tilde{X}}{d\xi}\frac{\partial \xi}{\partial r}\Big|_t$$

(8.27)

and

$$\frac{\partial X}{\partial t} = \tilde{X}(\xi)\frac{dX_1}{dt} + X_1 \frac{d\tilde{X}}{d\xi}\frac{\partial \xi}{\partial t}\Big|_r.$$

(8.28)

We then collect together the parts of the equations which are functions of ξ and t only, as usual for separation of variables problems.

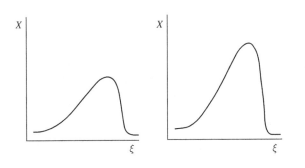

Fig. 8.4.

Note that in a similarity solution, a particular value of ξ does not necessarily comove with a particular fluid element. Instead, ξ labels a particular feature in the flow, so, for example, one value of ξ is associated at all times with the shock front. Therefore, straight from the definition of ξ, you find that

$$R_{\text{shock}} \propto \left(\frac{E}{\rho_0}\right)^{\frac{1}{5}} t^{\frac{2}{5}} \tag{8.29}$$

i.e. we have the same scaling as before.

8.2 Blast waves in astrophysics and elsewhere

How do we really know that blast waves behave in this self-similar way? Experimental verification has been provided by data on the expansion of the blast waves from atomic bombs. Not only does this confirm the scaling $r^{\frac{5}{2}} \propto t$, but also enables one to deduce the value of the explosion energy, E.

Figure 8.5 shows a series of snapshots of the first nuclear bomb test in New Mexico in 1945 and Figure 8.6 plots the radius of the explosion as a function of time.

There are two unknowns in this, the energy of the blast wave and the time t_0 when it started (which is not necessarily zero on the scale that was used). One can deduce that the clock started a little early (about 0.05 ms), and that the agreement with the $R \propto t^{\frac{2}{5}}$ model is very good.

In the case of supernova explosions, we can use the similarity solution to predict the radius of the swept up shell as a function of time and explosion energy. From equation (8.25), we can write

$$R(t) = \xi_0 \left(\frac{Et^2}{\rho_0}\right)^{\frac{1}{5}}. \tag{8.30}$$

where ξ_0 is the value of the similarity variable labeling the location of the shock front, which we assume to be of order unity.

The expansion velocity is obviously

$$u_S(t) = \frac{dR}{dt} = \frac{2}{5}\xi_0 \left(\frac{E}{\rho_0 t^3}\right)^{\frac{1}{5}} = \frac{2}{5}\frac{R}{t}. \tag{8.31}$$

In a 'typical' supernova explosion about 1 $M_\odot$ is ejected with a speed of about 10^4 km s^{-1}, so the energy of the explosion is about 10^{44} J. For a interstellar medium density of around 10^{-21} kg m^{-3}, we then have

$$R \sim 10^{13} t^{\frac{2}{5}} \quad \text{and} \quad u_S \sim 4 \times 10^{12} t^{-\frac{3}{5}}, \tag{8.32}$$

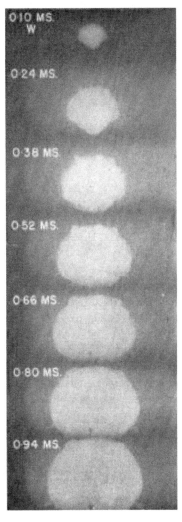

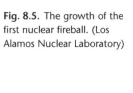

Fig. 8.5. The growth of the first nuclear fireball. (Los Alamos Nuclear Laboratory)

100 metres

or, if we express R in terms of parsecs $(3 \times 10^{16} \mathrm{m})$ and t in years $(3 \times 10^7 \mathrm{s})$

$$R \sim 0.3t^{\frac{2}{5}} \mathrm{pc} \quad \text{and} \quad u_S \sim 10^5 t^{-\frac{3}{5}} \mathrm{km \ s^{-1}}. \tag{8.33}$$

We know that this approximation is invalid for $t < 100$ years (since we know the expansion velocity is initially $10^4 \mathrm{km \ s^{-1}}$). Also, the derivation above requires that the mass of gas swept up from the interstellar medium by the blast wave is significantly greater than the

Fig. 8.6. Nuclear fireball size as a function of time.

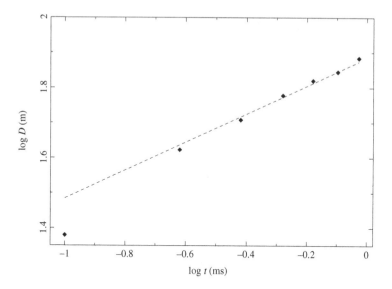

mass of the supernova ejecta. The timescale for this can be somewhat longer – observations of the remant of Tycho's supernova of 1572 suggest that it is yet to enter the Sedov phase. After a timescale of 10^5 years, energy losses become significant. Hence the adiabatic similarity solution only applies for a limited period: however, this time period is the one which is relevant to most supernova remnants.

Owing to the power law nature of the expansion, it is impossible to check the validity of equation (8.25) within observers' lifetimes unless the supernova exploded very recently, and recent ones are just those for which the Sedov solution does not yet apply! In any case, the clumpiness of the interstellar medium means that it is unlikely to follow the Sedov solution exactly, as the appearance of recent supernova remnants suggests (Figure 8.7).

8.3 Structure of the blast wave

What we have done so far is to have developed the description of a blast wave using some approximations, which we have justified with varying levels of conviction, and then written down a self-similar prescription and shown we obtained the same time-dependence for the blast wave radius. Now we take the self-similar nature of the blast wave, and investigate the nature of this self-similar solution more rigorously. Among other things, this will show us how good are the approximations we made when we developed the physical description.

To construct a similarity solution, every physical variable is written as a quantity with the correct dimensions constructed from r, t, E and

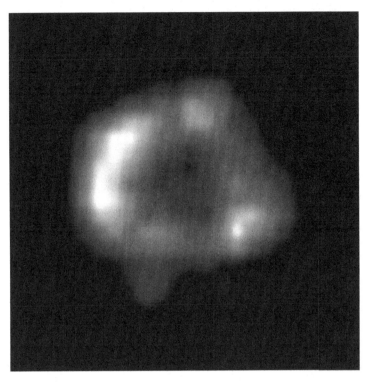

Fig. 8.7. An X-ray image of the blast wave from supernova 1987A taken about 12 years after the event. A shock wave, travelling at a speed of $4,500\,km\,s^{-1}$, is smashing into portions of a ring of material ejected by the star thousands of years earlier. (Chandra X-ray Observatory NASA/ CXC/SAO/PSU/D. Burrows *et al.*)

ρ_0 multiplied by a function of the dimensionless similarity variable ξ (8.25). Since $r = r(\xi, t)$ this then boils down to an expression of the form (8.26) which we have already discussed. As an example, consider $u(r, t)$ which we write as $A(r/t)u'(\xi)$ (for some constant A) or, equivalently, as $U(r/t)/(R/t))u'(\xi)$; here U is the velocity of the shocked gas in the frame of the explosion and R is the radius of the shock front, where we have normalised $u'(\xi_s) = 1$, with ξ_s being the value of the similarity variable at the shock front. Combining equations (8.6), (8.12), (8.25) and (8.31) we write

$$u = \frac{4}{5}(\gamma+1)\left(\frac{r}{t}\right)u'(\xi) = \frac{4}{5}(\gamma+1)\left(\frac{\lambda}{t}\right)\xi u'(\xi) \qquad (8.34)$$

where the latter is in the form (8.26). Similarly we can write

$$\rho = \rho_0\left(\frac{\gamma+1}{\gamma-1}\right)\rho'(\xi) \qquad (8.35)$$

where we have used (8.1) and again required $\rho'(\xi_s) = 1$. Likewise $p = B\rho_0(r/t)^2 p'(\xi)$ (for some constant B) which we write as

$$p = p_1 \left(\frac{(r/t)}{(R/t)} \right)^2 p'(\xi) \qquad (8.36)$$

and we use (8.9), (8.12) and (8.31) to write this as

$$p = \frac{8}{25} \frac{\rho_0}{(\gamma+1)} \left(\frac{r}{t} \right)^2 p'(\xi) = \frac{8}{25} \frac{\rho_0}{(\gamma+1)} \left(\frac{\lambda}{t} \right)^2 \xi^2 p'(\xi) \qquad (8.37)$$

where we also require $p'(\xi_s) = 1$ and note that (8.37) is of the form (8.26).

We can now proceed to substitute these expressions into the spherical continuity and momentum equations:

$$\frac{\partial \rho}{\partial t} + \frac{1}{r^2} \frac{\partial}{\partial r} (r^2 \rho u) = 0, \qquad (8.38)$$

$$\frac{\partial u}{\partial t} + u \frac{\partial u}{\partial r} = -\frac{1}{\rho} \frac{\partial p}{\partial r}, \qquad (8.39)$$

and the condition that for each fluid element in the shocked gas the entropy is constant (as appropriate for an adiabatic shock):

$$\left(\frac{\partial}{\partial t} + u \frac{\partial}{\partial r} \right) \log \frac{p}{\rho^\gamma} = 0. \qquad (8.40)$$

When substituting the forms for ρ, p and u derived above, we recall that the derivatives are transformed according to equations (8.27) and (8.28).

After a little manipulation, the continuity, momentum and entropy conservation equations become:

$$-\xi \frac{d\rho'}{d\xi} + \frac{2}{\gamma+1} \left[3\rho'u' + \xi \frac{d}{d\xi}(\rho'u') \right] = 0, \qquad (8.41)$$

$$-u' - \frac{2}{5} \xi \frac{du'}{d\xi} + \frac{4}{5(\gamma+1)} \left(u'^2 + u'\xi \frac{du'}{d\xi} \right) = -\frac{2(\gamma-1)}{5(\gamma+1)} \frac{1}{\rho'} \left(2p' + \xi \frac{dp'}{d\xi} \right), \qquad (8.42)$$

and

$$\xi \frac{d}{d\xi} \left(\log \frac{p'}{\rho'^\gamma} \right) = \frac{5(\gamma+1) - 4u'}{2u' - (\gamma+1)}. \qquad (8.43)$$

The important thing here is not the details of the resultant equations, but the fact that r and t have cancelled out from the Eulerian equations

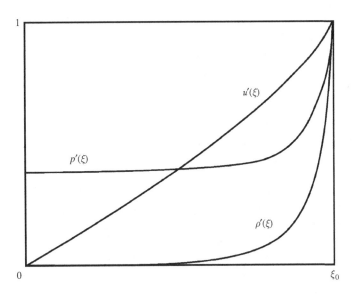

Fig. 8.8.

expressed in this form, leaving ξ as the only independent variable. So we have established that self-similar solutions are possible using the dimensionless variables as we have defined them. Now all we have to do is solve the coupled set of equations (8.41)–(8.43) subject to the boundary conditions $\rho'(\xi_0) = u'(\xi_0) = p'(\xi_0) = 1$.

The boundary conditions apply at ξ_0 though, which we still have to determine. We do this by using the fact that the total energy of the blast wave remains a constant. So

$$E = \int_0^R \left(\frac{\rho u^2}{2} + \frac{p}{\gamma - 1} \right) 4\pi r^2 \, dr \qquad (8.44)$$

has to be translated into dimensionless terms. This becomes

$$\frac{32\pi}{25(\gamma^2 - 1)} \int_0^{\xi_0} \left[p' + \rho' u'^2 \right] \xi^4 \, d\xi = 1, \qquad (8.45)$$

so ξ_0 can be found.

The integration of these equations is described as 'elementary but laborious' (Landau & Lifshitz, *Fluid Mechanics* in §116, who give the solution found by Sedov in 1946) once you have set up variables so that the integral (8.45) is known. It is probably more enlightening to show what it looks like (Figure 8.8 is for $\gamma = 7/5$).

You can now see whether or not the assumptions made in establishing the physical picture are valid. Much of the swept-up mass is in the shell just behind the shock, and the idea that the post-shock pressure is some multiple of the cavity pressure also looks acceptable.

The single velocity applying to the shell does not look quite as good, but some suitably weighted average will be acceptable.

8.4 Breakdown of the similarity solution

The similarity solution works well when there is no significant counter-pressure from outside the shock, but the feature of the equations which allows it (effectively being able to put everything in scale-free post-shock variables) breaks down when $p_0 = \frac{\mathcal{R}_*}{\mu} \rho_0 T_0$ becomes significant, i.e. when the pressure in the shell $p_1 \sim p_0$. Since $p_1 \sim \rho_0 u_0^2$, and the sound speed is $\frac{p}{\rho}$ (to within a factor γ in any case), then this condition is equivalent to $u_0 \sim c_s$, i.e. that the shell is no longer moving super-sonically with respect to the medium. Under these circumstances the blast wave character has gone, since now a sound wave can propagate ahead of the residual motion of the blast wave. So the picture we had in the blast wave phase (left in Figure 8.9) has evolved into something rather different (right).

During the blast wave phase all the gas in its path is swept up in to a shell and after that moves with the shell. In the late stages the disturbance passes in to the undisturbed gas as a mild compression followed by a rarefaction, and after the wave has passed the gas returns to its original state. (Any reader who is familiar with surfing will be able to relate this distinction to what happens when a wave breaks. The breaking wave is like the blast wave and the surfer is carried along as part of the bulk transport of water in the path of the wave. Once the wave has broken, however, a surfer in the shallows will be 'informed' of the disturbance further out to sea by the arrival of ripples on which the board rises and falls but is not transported). In the case of the supernova, the radius which the blast wave reaches at the point that it makes the transition between blast wave and sonic wave behaviour gives the maximum size of a bubble blown in the medium.

We can estimate the maximum blast wave radius, R_{max} by setting $p_1 \sim p_0 \sim \rho_0 c_0^2/\gamma$, where c_0 is the sound speed in the undisturbed medium, which implies that $u_0^2 \sim \frac{(\gamma+1)}{2\gamma} c_0^2$ (using equation 8.9). But we can obtain u_0 as a function of R from the equation we have for the

Fig. 8.9. Left, blast wave phase; right, late phase.

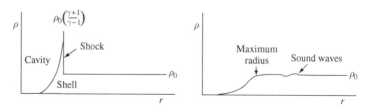

energy (8.20), which was $E = \frac{4\pi}{3} R^3 \left[\frac{1}{2}\rho \left(\frac{2u_0}{\gamma+1} \right)^2 + \frac{\alpha}{\gamma-1} \frac{2\rho_0 u_0^2}{\gamma+1} \right]$, provided we choose α. The Sedov solution shows us that $\alpha \sim \frac{1}{2}$, so we find

$$u_0^2 = \frac{(\gamma+1)(\gamma^2-1)}{(3\gamma-1)} \frac{3E}{4\pi\rho_0 R^3}. \tag{8.46}$$

Combining these we have

$$E \sim \frac{3\gamma-1}{2(\gamma+1)} \frac{4\pi}{3} \rho_0 R_{max}^3 \frac{c_0^2}{\gamma(\gamma-1)}. \tag{8.47}$$

The $\frac{4\pi}{3}\rho_0 R^3 \frac{c_0^2}{\gamma(\gamma-1)}$ term is the thermal energy originally contained within a sphere of radius R_{max}, since $\frac{c_0^2}{\gamma(\gamma-1)} = \frac{p_0}{\gamma-1}$ is the thermal energy per unit volume in the undisturbed gas.

So the condition $p_1 \sim p_0$ (which is approximately the criterion $u_0 = c_0$) is also approximately the criterion that the blast wave gets out to a point where the *explosion energy is equal to the total thermal energy contained within that sphere*. [Note: all of these criteria are approximate because there is no sudden transition between blast wave and sound wave behaviour.]

We could have arrived at this conclusion through dimensional arguments almost as well. The similarity solution worked because if E and ρ_0 are the only two parameters of the problem then we cannot construct characteristic length- and time-scales from these. If we introduce a finite temperature in the undisturbed region, so $c_0 \neq 0$, then we can construct a lengthscale $r \sim \left(\frac{E}{\rho_0 c_0^2} \right)^{\frac{1}{3}}$ and timescale $t \sim \frac{1}{c_0} \left(\frac{E}{\rho_0 c_0^2} \right)^{\frac{1}{3}}$. This is the lengthscale where the similarity solution ceases to apply.

We are now in a position to quantify the characteristic scale, R_{max}, for the maximum size attained by a supernova driven bubble in the ISM. The timescale on which the bubble attains this size (t_{max}) is roughly the timescale on which a sound wave in the undisturbed ISM would cross R_{max} (i.e. R_{max}/c_0), which is also roughly the timescale on which the fully blown bubble is enroached by the ISM and dissolves. Typical values for the density and temperature of the atomic phase of the ISM are $T \sim 10^4$K and $\rho \sim 10^{-21}$ kg m^{-3}. In this case R_{max} turns out to be several 100 pc and t_{max} roughly 10 Myr. Now the supernova rate in the Milky Way is roughly 10^{-7} Myr^{-1} per cubic parsec, so within a time t_{max}, the volume of ISM containing one supernova on average is $\sim 10^6$ pc^3. However, the expected final bubble volume ($4\pi R_{max}^3/3$) exceeds this by about two orders of magnitude. Therefore we should conclude from this that the filling factor of supernova driven bubbles should be much larger than unity - in other words, the entire

ISM should be permanently heated to high temperatures ($> 10^6$ K) by supernova explosions. In reality, this is clearly not the case (in other words, the ISM contains significant components that are much cooler than this). We explore below the reasons that supernovae are in fact much less efficient at heating the ISM than the above estimate suggests.

8.5 The effects of cooling and blowout from galactic disks

In all the above analysis, we have assumed that the bubble evolves adiabatically, in other words that we can neglect radiative losses from the shocked gas. However, if one includes radiative cooling (see Chapter 4), one finds that after $\sim 10^5$ years (when the bubble has grown to about 20 pc), the gas behind the shock front is able to cool significantly and collapses to a much thinner shell than that shown for the adiabatic solution in Figure 8.8. The bubble now grows more slowly than the adiabatic solution ($R \propto t^{2/5}$), due to radiative energy loss from the system. In the case that the bubble interior is able to cool completely, the shell would no longer be driven and thus would follow a momentum conserving evolution: one may readily demonstrate that since, in this case, $\dot{R}R^3$ would be constant, the power law evolution of the bubble radius would be $R \propto t^{1/4}$. In practice, hydrodynamic calculations with a realistic cooling law show that, once cooling becomes important, the bubble expands according to a law ($R \propto t^{0.3}$) that is intermediate between those appropriate to adiabatic and momentum conserving evolution. The bubble expansion speed falls more rapidly than in the adiabatic case and attains a value that is comparable with the sound speed in the undisturbed medium after about a Myr, when the bubble is only about 50 pc in radius. This radius is smaller than that in the adiabatic case by a factor of a few; when cubed, this results in the filling factor of supernova bubbles being about two orders of magnitude less than in the adiabatic case. Since the counter-pressure of the ISM on the bubble again becomes important at the point that the thermal energy of the original ISM within the bubble volume is comparable with the input of supernova energy into this volume, we can also immediately deduce that the amount of supernova energy that is *actually deposited in the ISM* (as opposed to being radiated away) is only of order a per cent or so of the total explosion energy.

Another factor that also diminishes the coupling between supernova energy and the ISM is the finite scale height of the gas in a direction perpendicular to the galactic disk. All the above solutions assumed that the supernova exploded into an infinite and uniform medium. If there

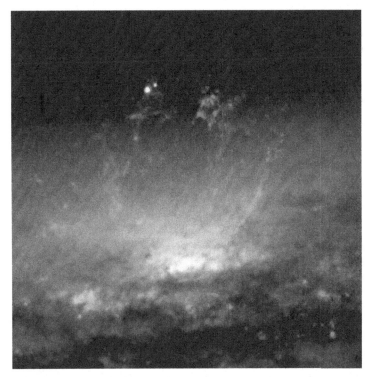

Fig. 8.10. The nuclear regions of NGC 3079, where a lumpy bubble of hot gas is rising from a cauldron of glowing matter. The bubble is more than 1 kpc wide and rises a similar distance above the galactic disc. The filamentary structure is seen in hydrogen $H\alpha$ line emission. (NASA/STScI)

is instead a steep density gradient (as occurs as the bubble propagates away from the galactic mid-plane), then the diminishing density ahead of the shock front causes it to accelerate and the bubble 'blows out' of the plane, depositing most of its into energy into the low density halo of the galaxy. (Once the bubble is de-pressurised by the blow out of hot gas into the halo, the bubble expansion in the plane is no longer driven and it therefore expands thereafter in a momentum conserving fashion, until its velocity approaches the sound speed of the ISM). The typical size scale achieved by bubbles as a result of galactic blow out is again of order 100 pc; this implies that (even in the absence of radiative cooling) the finite scale height of disk galaxies causes the bubble filling factor to be considerably less than the estimate made above for adiabatic evolution in an infinte medium.

The net effect of both cooling and galactic blowout is thus that the fraction of the ISM in the disk that is filled with supernova driven bubbles is somewhat less than unity, and that the fraction of the energy of a supernova that is deposited in the ISM of the galactic disk is only about a per cent of the total explosion energy. Depending on the details of the cooling, howver, considerable energy may be deposited in the galactic halo. The combined effects of multiple supernovae in disk galaxies (particularly those undergoing spectacular bursts of star formation in their nuclei) is thus to create flows of hot gas out of the disk plane that have variously been termed galactic superwinds or galactic fountains, as well as channels in the disk known as galactic chimneys. Figure 8.10 demonstrates evidence of such flows in the nucleus of NGC3079.

Chapter 9
Bernoulli's equation

Bernoulli's equation is of central importance to fluid dynamics in both terrestrial and astronomical applications. Indeed, as is well known, it provides the explanation of how aeroplanes get off the ground. In astronomy we will apply it to diverse situations where we have steady, barotropic flows, from the case of accretion of gas onto stars to the study of the collimated outflows that emanate from young stars and galactic nuclei.

9.1 Basic equation

We start with the time-dependent momentum equation:

$$\frac{\partial \mathbf{u}}{\partial t} + \mathbf{u} \cdot \nabla \mathbf{u} = -\frac{1}{\rho} \nabla p - \nabla \Psi. \tag{9.1}$$

Now if we have a barotropic equation of state (i.e. $p = p(\rho)$), then we can write

$$\frac{1}{\rho} \nabla p = \nabla \int \frac{\mathrm{d}p}{\rho}. \tag{9.2}$$

We also know

$$\mathbf{u} \cdot \nabla \mathbf{u} = \nabla \left(\frac{1}{2} u^2 \right) - \mathbf{u} \wedge (\nabla \wedge \mathbf{u}), \tag{9.3}$$

which can be compactly derived in summation convention. $\mathbf{u} \wedge (\nabla \wedge \mathbf{u})$ becomes, in component terms

$$\epsilon_{ijk} u_j \epsilon_{klm} \partial_l u_m = \epsilon_{ijk} \epsilon_{klm} u_j \partial_l u_m$$

$$= (\delta_{il} \delta_{jm} - \delta_{jl} \delta_{im}) u_j \partial_l u_m$$

$$= u_j \partial_i u_j - u_j \partial_j u_i.$$

Translating this back to vector notation gives

$$\mathbf{u} \wedge (\boldsymbol{\nabla} \wedge \mathbf{u}) = \boldsymbol{\nabla} \left(\frac{1}{2} u^2 \right) - \mathbf{u} \cdot \boldsymbol{\nabla} \mathbf{u}. \tag{9.4}$$

Now defining the *vorticity* **w** by

$$\mathbf{w} = \boldsymbol{\nabla} \wedge \mathbf{u}, \tag{9.5}$$

(9.1) becomes

$$\frac{\partial \mathbf{u}}{\partial t} + \boldsymbol{\nabla} \left(\frac{1}{2} u^2 \right) - \mathbf{u} \wedge \mathbf{w} = -\boldsymbol{\nabla} \left[\int \frac{\mathrm{d}p}{\rho} + \Psi \right]. \tag{9.6}$$

Taking the dot product of this equation with **u**, it follows that, *if the flow is steady* (so $\frac{\partial \mathbf{u}}{\partial t} = 0$), then

$$\mathbf{u} \cdot \boldsymbol{\nabla} \left[\frac{1}{2} u^2 + \int \frac{\mathrm{d}p}{\rho} + \Psi \right] = 0. \tag{9.7}$$

The term containing the vorticity **w** has disappeared because of the identity $\mathbf{u} \cdot (\mathbf{u} \wedge \mathbf{w}) = 0$ (the cross product of **u** with anything is perpendicular to **u** and hence the dot product of this vector with **u** is necessarily zero).

Therefore for a steady barotropic flow, the quantity

$$H = \frac{1}{2} u^2 + \int \frac{\mathrm{d}p}{\rho} + \Psi \tag{9.8}$$

is constant along streamlines, and is referred to as Bernoulli's constant.

For a physical interpretation of the constancy of H along streamlines we first consider the $p = 0$ case. Then the constancy of H just implies that the kinetic energy plus potential energy along streamlines is a constant.

For a fluid flow with non-zero pressure, in kinetic theory Bernoulli's constant represents the conversion of kinetic energy between random molecular motions and bulk flow, as illustrated in Figure 9.1.

In hydrodynamic terms, we can understand the constancy of H as reflecting the fact that pressure differences are required to accelerate the flow, as illustrated in Figure 9.2.

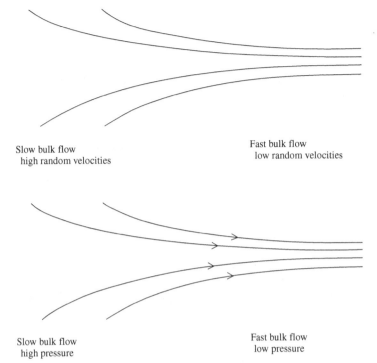

Fig. 9.1. Bernoulli's constant – kinetic theory interpretation.

Slow bulk flow
high random velocities

Fast bulk flow
low random velocities

Fig. 9.2. Bernoulli's constant – hydrodynamic interpretation.

Slow bulk flow
high pressure

Fast bulk flow
low pressure

Everyday examples of Bernoulli's equation at work include aeroplanes (the wing is curved, so the airspeed over the top is higher and the pressure lower, so there is a lift force due to the pressure differential). Another less well-known manifestation of Bernoulli's equation is what happens when you turn on the shower and in consequence the shower curtain bows in and tries to squeeze you against the wall. This is because the downward flowing water accelerates the adjoining air as well, so that there is a region of downward flowing air inside the shower curtain. In consequence, the air pressure drops inside the shower cabin and the curtain is pushed in by the external air pressure.

If the flow is steady and in addition $\mathbf{w} = 0$ (i.e. the flow is 'curl free', or 'irrotational') then

$$\nabla H = 0, \tag{9.9}$$

which implies that H is constant *everywhere*.

What is the nature of curl-free flow? Stokes' theorem states that

$$\oint \mathbf{u} \cdot d\boldsymbol{\ell} = \int \nabla \wedge \mathbf{u} \cdot d\mathbf{S}, \tag{9.10}$$

so

$$\mathbf{V} \wedge \mathbf{u} = 0 \qquad \Rightarrow \qquad \oint \mathbf{u} \cdot d\boldsymbol{\ell} = 0. \qquad (9.11)$$

This is satisfied e.g. by a uniform flow, but is not satisfied by general rotating flows $\mathbf{u} = u(R)\hat{\mathbf{e}}_\phi$ (here in cylindrical polars). This is why the case $\mathbf{w} = 0$ is termed 'irrotational'.

This is however slightly misleading, since there is one class of rotating flow for which $\mathbf{V} \wedge \mathbf{u} = 0$, and that is flows of the form $\mathbf{u} \propto \frac{1}{R}\hat{\mathbf{e}}_\phi$. Then $\mathbf{V} \wedge \mathbf{u} = \frac{1}{R}\frac{\partial}{\partial R} R u(R)\hat{\mathbf{e}}_z = 0$, but $\oint \mathbf{u} \cdot d\boldsymbol{\ell} \neq 0$. The reason for this apparent violation of Stokes' theorem is that there is a singularity at $R = 0$, and the theorem does not apply if there are singularities within the surface of integration. You can choose a different contour for the integration which does not include the singularity, and then, of course, the theorem does apply (see Figure 9.3).

In a general flow we have

$$\frac{\partial \mathbf{u}}{\partial t} = -\mathbf{V}H + \mathbf{u} \wedge \mathbf{w}. \qquad (9.12)$$

If we take the curl of both sides we have

$$\frac{\partial}{\partial t}\mathbf{V} \wedge \mathbf{u} = -\mathbf{V} \wedge \mathbf{V}H + \mathbf{V} \wedge (\mathbf{u} \wedge \mathbf{w}). \qquad (9.13)$$

$\mathbf{V} \wedge \mathbf{V}H = 0$ from the 'curl grad is zero' identity, so

$$\frac{\partial \mathbf{w}}{\partial t} = \mathbf{V} \wedge (\mathbf{u} \wedge \mathbf{w}). \qquad (9.14)$$

The importance of this result, known as *Helmholtz's equation*, is that if $\mathbf{w} = 0$ initially, then it stays zero. (If we had however

Fig. 9.3. Stokes' theorem.

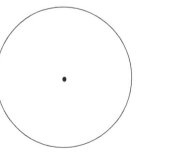

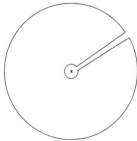

Not Applicable because
of singularity at centre

Applicable because surface
includes no singularities

included viscous terms in the momentum equation this would no longer be true and vorticity could in this case be created *ab initio*: see Chapter 11.)

There is a more general result which comes from this as well. If we have a surface S inside a fluid, then it turns out that the flux of vorticity linked with this surface

$$\int_S \mathbf{w} \cdot d\mathbf{S} \tag{9.15}$$

is a constant which moves with the fluid, i.e. the Lagrangian derivative

$$\frac{D}{Dt} \int_S \mathbf{w} \cdot d\mathbf{S} = 0. \tag{9.16}$$

In order to show this, note that $\int_S \mathbf{w} \cdot d\mathbf{S}$ can change with time for two reasons – the intrinsic change in $\mathbf{w}$, and the change in the surface S caused by the flow, so

$$\frac{D}{Dt} \int_S \mathbf{w} \cdot d\mathbf{S} = \int_S \frac{\partial \mathbf{w}}{\partial t} \cdot d\mathbf{S} + \int_S \mathbf{w} \cdot \frac{D\,d\mathbf{S}}{Dt}. \tag{9.17}$$

Figure 9.4 shows an element of area which has changed from $d\mathbf{S}$ to $d\mathbf{S}'$ in a time interval δt. The vector area of the sides of the volume which has these elements as ends is $-\delta t\,\mathbf{u} \wedge \delta \mathbf{l}$, where $\delta \mathbf{l}$ is a length element on the curve bounding $d\mathbf{S}$. Since the vector area over an entire volume $\int d\mathbf{S} = 0$, we have

$$d\mathbf{S}' - d\mathbf{S} - \delta t \oint \mathbf{u} \wedge d\mathbf{l} = 0, \tag{9.18}$$

Fig. 9.4.

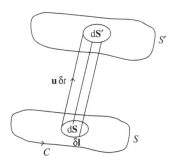

where the line integral is around the surface dS. From this it follows that

$$\frac{D\,dS}{Dt} = \oint \mathbf{u} \wedge d\mathbf{l}. \tag{9.19}$$

So, over the surface S, we have

$$\int_S \mathbf{w} \cdot \frac{D\,dS}{Dt} = \int \oint \mathbf{w} \cdot (\mathbf{u} \wedge \, d\mathbf{l}) = \int \oint \mathbf{w} \wedge \mathbf{u} \cdot d\mathbf{l}. \tag{9.20}$$

The double integral means that we first take the line integral around dS and then integrate to make up the surface S. Then the total line integral is around C which encircles the whole surface S, since inner components cancel out. Then

$$\int_S \mathbf{w} \cdot \frac{D\,dS}{Dt} = \oint_C \mathbf{w} \wedge \mathbf{u} \cdot d\mathbf{l} = \int_S \boldsymbol{\nabla} \wedge (\mathbf{w} \wedge \mathbf{u}) \cdot dS, \tag{9.21}$$

by Stokes' theorem. Therefore (since $\mathbf{u} \wedge \mathbf{w} = -\mathbf{w} \wedge \mathbf{u}$)

$$\frac{D}{Dt} \int_S \mathbf{w} \cdot dS = \int_S dS \cdot \left[\frac{\partial \mathbf{w}}{\partial t} - \boldsymbol{\nabla} \wedge (\mathbf{u} \wedge \mathbf{w}) \right]. \tag{9.22}$$

The right hand side is zero from Helmholtz's equation, so *the flux of the vorticity* $\mathbf{w}$ *is conserved and moves with the fluid*. This is referred to as Kelvin's vorticity theorem.

If we have a flow which has zero vorticity, so $\boldsymbol{\nabla} \wedge \mathbf{u} = 0$, then we can write the velocity as the gradient of some potential function Φ_u, i.e. there exists a Φ_u such that

$$u = -\boldsymbol{\nabla}\Phi_u. \tag{9.23}$$

If the flow is incompressible, then $\boldsymbol{\nabla} \cdot \mathbf{u} = 0$, and so for flows that are *both incompressible and irrotational*:

$$\nabla^2 \Phi_u = 0. \tag{9.24}$$

This is Laplace's equation, and is the same equation as for a gravitational potential in a vacuum (though with rather different boundary conditions generally!). Therefore some of the techniques that are familiar from solving the structure of gravitational/electrostatic fields in vacuo (e.g. the method of images) may also be applicable in the case of incompressible fluids. This result is therefore of great applicability in terrestrial fluid mechanics, where incompressibility is a reasonable approximation, but of rather limited use in most astrophysical applications. See however its application to the Rayleigh–Taylor instability in Chapter 10.

Another result which is useful for two-dimensional flows of an incompressible fluid results from the fact that, if there is no z-dependence, then we can write

$$\mathbf{u} = -\nabla \wedge [\varphi(x, y)\hat{\mathbf{e}}_z]. \tag{9.25}$$

The reason why we can write it in this form is that we can invoke the identity 'div of curl is zero', so that incompressible flows (i.e. with $\nabla \cdot \mathbf{u} = 0$) must be expressible as the curl of something. On the other hand, in order to ensure that the vector $\mathbf{u}$ only has components in the (x,y) plane, which depend only on x and y, the 'something' must be of the form shown in (9.25). Thus

$$u_x = -\frac{\partial \varphi}{\partial y} \tag{9.26}$$

and

$$u_y = \frac{\partial \varphi}{\partial x} \tag{9.27}$$

and it can be easily verified that $\nabla \cdot \mathbf{u} = 0$. The change in φ along a streamline is $\frac{\partial \varphi}{\partial x}dx + \frac{\partial \varphi}{\partial y}dy$ which, from (9.26) and (9.27), is zero. Thus the function φ is a stream function (i.e. constant along streamlines).

9.2 De Laval nozzle

Consider a steady flow in the z direction in a tube of given variable cross-section $A(z)$. Since the flow is steady, we have directly from mass conservation

$$\rho u A = \text{constant} = \dot{M} \text{ (mass flow per second).} \tag{9.28}$$

The momentum equation for a steady flow with no gravity is

$$\mathbf{u} \cdot \nabla \mathbf{u} = -\frac{1}{\rho}\nabla p = -\frac{1}{\rho}\nabla \rho \frac{dp}{d\rho}, \tag{9.29}$$

where in the last term we have assumed that the equation of state is barotropic. From (9.28) we have

$$\ln \rho + \ln u + \ln A = \ln \dot{M}, \tag{9.30}$$

$\Rightarrow$

$$\frac{1}{\rho}\boldsymbol{\nabla}\rho = -\boldsymbol{\nabla}\ln u - \boldsymbol{\nabla}\ln A. \tag{9.31}$$

Therefore (9.29) becomes

$$\mathbf{u} \cdot \boldsymbol{\nabla}\mathbf{u} = [\boldsymbol{\nabla}\ln u + \boldsymbol{\nabla}\ln A]\frac{\mathrm{d}p}{\mathrm{d}\rho}. \tag{9.32}$$

Now $\frac{\mathrm{d}p}{\mathrm{d}\rho} = c_\mathrm{s}^2$, where c_s is the sound speed, and if the flow is irrotational $\mathbf{u} \cdot \boldsymbol{\nabla}\mathbf{u} = \boldsymbol{\nabla}(\frac{1}{2}u^2) = u\boldsymbol{\nabla}u = u^2\boldsymbol{\nabla}\ln u$, so (9.32) becomes

$$\left(u^2 - c_\mathrm{s}^2\right)\boldsymbol{\nabla}\ln u = c_\mathrm{s}^2\boldsymbol{\nabla}\ln A. \tag{9.33}$$

What this tells us is that a minimum or maximum in A corresponds to either

- a minimum or maximum in u, or
- $u = c_\mathrm{s}$.

Conversely, the gas can only make a sonic transition (i.e. subsonic to supersonic flow, or vice versa) at a maximum or minimum of the cross-sectional area of the nozzle.

Now we may apply Bernoulli's equation to the problem. Since there is no gravity, and the flow is steady and irrotational, we have

$$\frac{1}{2}u^2 + \int \frac{\mathrm{d}p}{\rho} = \text{constant.} \tag{9.34}$$

If we are to do anything with this, we need to know the equation of state, so we will try the standard ones in turn.

I Isothermal
If $p = \frac{\mathcal{R}_*}{\mu}\rho T$ for constant T, then

$$\int \frac{\mathrm{d}p}{\rho} = \frac{\mathcal{R}_*}{\mu}T\ln\rho = c_\mathrm{s}^2\ln\rho. \tag{9.35}$$

From (9.33) we know that *if* there is a transition between subsonic and supersonic flow (or vice versa) then it *must* occur at a maximum or minimum of A ($A = A_m$ say) in the pipe, i.e. $u|_{A_m} = c_s$. (Note that it is also possible for the flow to remain subsonic or supersonic throughout, in which case, for steady flow, u must attain an extremum at the extremum of A. Here however we examine the solutions where the flow instead makes a sonic transition at $A = A_m$.) Therefore if $\dot{M}$ is specified, and we know c_s everywhere (because the flow is isothermal at some specified temperature) and we also know the cross-section area $A(z)$, then from Equation (9.28) we know $\rho|_{A_m}$. We can then find the run of ρ everywhere using (9.28) and (9.34). Equation (9.34) implies

$$\frac{1}{2}u^2 + c_s^2 \ln \rho = \frac{1}{2}c_s^2 + c_s^2 \ln \rho|_{A_m}, \tag{9.36}$$

so

$$u^2 = c_s^2 \left[1 + 2\ln \left(\frac{\rho|_{A_m}}{\rho} \right) \right]. \tag{9.37}$$

From (9.28), $\frac{\rho|_{A_m}}{\rho} = \frac{uA}{c_s A_m}$, so

$$u^2 = c_s^2 \left[1 + 2\ln \left(\frac{uA}{c_s A_m} \right) \right]. \tag{9.38}$$

Therefore, given $A(z)$, we are in a position to determine $u(z)$ and hence $\rho(z)$, i.e. the structure of the flow everywhere subject to a given $\dot{M}$ and c_s.

II Polytropic (and adiabatic)
We now consider a more general barotropic equation of state which can be parameterised by the polytropic relation $p = K\rho^{1+\frac{1}{n}}$. (Note that this does not imply $p = K\rho^\gamma$ unless in addition the flow is isentropic.)

We can still use the fact that the flow makes a sonic transition at $A = A_m$ but the problem is a bit more complex than the isothermal case because we do not know c_s (at the sonic point or anywhere else) a priori because c_s is a function of the density, which is a quantity for which we are solving. Specifically, we have

$$c_s^2 = \left(\frac{n+1}{n} \right) K\rho^{\frac{1}{n}} \tag{9.39}$$

and

$$\int \frac{\mathrm{d}p}{\rho} = nc_s^2. \tag{9.40}$$

From (9.28) we have

$$\rho|_{A_m} c_s|_{A_m} A_m = \dot{M}, \tag{9.41}$$

so

$$\rho|_{A_m} \left(\frac{n+1}{n}\right)^{\frac{1}{2}} K^{\frac{1}{2}} \rho|_{A_m}^{\frac{1}{2n}} A_m = \dot{M}. \tag{9.42}$$

This gives

$$\rho|_{A_m} = \left[\left(\frac{\dot{M}}{A_m}\right)^2 \frac{n}{K(n+1)}\right]^{\frac{n}{2n+1}}, \tag{9.43}$$

and so $c_s|_{A_m}$ and $u|_{A_m}$ can be determined. Then using Bernoulli's equation, the run of variables with z follows, e.g.

$$\frac{1}{2}\left(\frac{\dot{M}}{A\rho}\right)^2 + (n+1)K\rho^{\frac{1}{n}} = \left(n+\frac{1}{2}\right)\left(\frac{n+1}{n}\right)K\rho_{A_m}^{\frac{1}{n}}. \tag{9.44}$$

De Laval nozzles: general points
From Equation (9.33)

$$\left(u^2 - c_s^2\right)\nabla \ln u = c_s^2 \nabla \ln A \tag{9.45}$$

we see that:

- In a subsonic regime ($u < c_s$), if A decreases then $\nabla \ln u$ is positive, and hence the speed u increases. This is the situation that we are used to from experience of incompressible flows (like rivers flowing through narrows).
- In the supersonic regime ($u > c_s$), then an increase in A results in an *increase* in u. What happens in this regime is that the reduction in the density is large, so the condition that $\dot{M}$ is a constant requires an increase of u as well as A. This slightly counter-intuitive result stems from the much greater compressibility of supersonic flows.

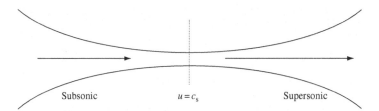

Subsonic $u = c_{\mathrm{s}}$ Supersonic

Fig. 9.5. De Laval nozzle.

Consequently a nozzle that gets progressively narrower and then wider can be used to accelerate a flow from a subsonic to a supersonic regime and u increases monotonically.

Evidently this result can be (and is) exploited in engineering applications (such as the design of jet engines) where one is free to construct pipes of any cross-section to achieve a desired flow pattern. It is less obvious, however, how this could be applicable to astronomy, i.e. what are the analogues of pipes in the interstellar medium.

One suggested application is to the narrow jets that are seen emanating from the centres of some galaxies (normally at radio wavelengths), which can be highly supersonic. It has been suggested that hot plasma, somehow generated in the 'central engines' of active galactic nuclei, is able to escape most easily in a direction perpendicular to the disc of gas in the centre of the galaxy. The gas disc therefore acts like the pipe in the above example, except that rather than being a solid wall, it is itself a fluid which confines the jet through providing a pressure balance boundary condition. Such an interpretation is supported by the more detailed images that are possible in the context of the relatively nearby jets that emanate from some young stars. Figure 9.6 depicts a collection of jets from young stars imaged with the Hubble Space Telescope: the disc around the star is clearly visible in the left hand image as the dark dust lane perpendicular to the jet. (The saucer-shaped emission on either side is interpreted as scattering of starlight off the surface of the disc.) The beaded structure in the right hand image illustrates an important respect in which real jets differ from the steady state solutions we have developed here: this structure is believed to result from time-dependent injection of material into the base of the flow region. As parcels of gas travelling at different speeds overtake each other, shocks are produced within the jet (in contrast to the shock-free nature of the steady state solution). The lower image contains further emission from shocks as the jet ploughs into the ambient medium.

Although such jets have been modelled as de Laval nozzles, there is a fundamental difficulty with this picture, since such purely thermal acceleration (i.e. acceleration through pressure gradients) implies such

Fig. 9.6. Jets from young stars, as imaged by WFPC2 on the Hubble Space Telescope. The scale bar at the bottom left of each panel represents 1000 au. (NASA/STScI)

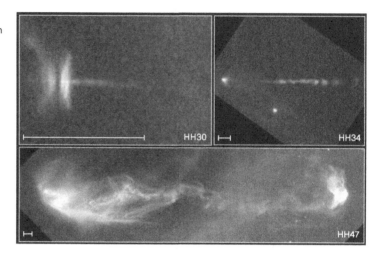

hot and dense conditions in the centre of the galaxy that these regions should cool down very quickly and cease to drive the jet activity. The picture must therefore be more complicated than this and – although pressure gradients and the confining interaction of the disc-like medium undoubtedly play a role – additional processes (probably involving magnetic fields) are probably involved in the acceleration and collimation of radio jets.

We can gain some further insight into the different compressibility of subsonic and supersonic flows from Equation (9.29):

$$\mathbf{u} \cdot \nabla \mathbf{u} = u^2 \nabla \ln u = -c_s^2 \nabla \ln \rho. \qquad (9.46)$$

- If $u \ll c_s$ then $\nabla \ln u \gg \nabla \ln \rho$. This implies that accelerations in the flow are important, and pressure changes small. Thus the fluid motion is nearly incompressible – and this is the reason that incompressible flows are normally a good description in terrestrial applications.
- If $u \gg c_s$ then $\nabla \ln u \ll \nabla \ln \rho$. Hence u is approximately constant and pressure gradients are not very important in accelerating the flow, but give rise to density changes predominantly.

9.3 Spherical accretion and winds

Another situation which we can study is the spherically symmetric accretion of gas onto a star, which we consider as a point mass. The gas is initially at rest at ∞, and so its inflow is subsonic at large distances from the star (so consequently ∇p is important). By the time it nears the star it may be essentially in free fall (i.e. supersonic flow in which ∇p is unimportant). If this is to happen, the gas has to make a sonic transition somewhere in between. In many respects, the problem is similar to the de Laval nozzle where we use Bernoulli's equation plus the continuity equation, with the added complication here that gravity is present.

In the steady state, the continuity equation becomes

$$\dot{M} = 4\pi r^2 \rho u \tag{9.47}$$

and the momentum equation for a steady flow is

$$u\frac{du}{dr} = -\frac{1}{\rho}\frac{dp}{dr} - \frac{GM}{r^2} \tag{9.48}$$

$\Rightarrow$

$$u^2\frac{d\ln u}{dr} = -c_s^2\frac{d\ln\rho}{dr} - \frac{GM}{r^2}. \tag{9.49}$$

From the continuity equation we have

$$\frac{d\ln\rho}{dr} = -\frac{2}{r} - \frac{d\ln u}{dr}, \tag{9.50}$$

$\Rightarrow$

$$\left(u^2 - c_s^2\right)\frac{d\ln u}{dr} = \frac{2c_s^2}{r}\left[1 - \frac{GM}{2c_s^2 r}\right]. \tag{9.51}$$

Therefore there is a radius in the flow

$$r_s \equiv \frac{GM}{2c_s^2} \tag{9.52}$$

such that *either* u is a maximum or minimum there *or else* $u = c_s$ there. Thus *if* a sonic transition occurs, it *must* do so at r_s. As before, further progress requires that we specify the equation of state.

I Isothermal case (Bondi accretion)
In this case, c_s is a constant, so we immediately know r_s if we know the temperature of the gas. Hence, from the continuity relation, we also know $\dot{M} = 4\pi r_s^2 c_s \rho_s$. However, in general we do not know either $\dot{M}$

or ρ_s a priori but instead have a boundary condition on the density at infinity. In order to relate this to ρ_s we need to examine the radial structure of the flow using Bernoulli's equation:

$$\frac{1}{2}u^2 + c_s^2 \ln\rho - \frac{GM}{r} = \frac{1}{2}c_s^2 + c_s^2 \ln\rho_s - \frac{GM}{r_s}. \tag{9.53}$$

Using $\frac{GM}{r_s} = 2c_s^2$ from above, this becomes

$$\frac{1}{2}u^2 + c_s^2 \ln\rho - \frac{GM}{r} = c_s^2\left(\ln\rho_s - \frac{3}{2}\right), \tag{9.54}$$

and so

$$u^2 = 2c_s^2\left[\ln\left(\frac{\rho_s}{\rho}\right) - \frac{3}{2}\right] + \frac{2GM}{r}. \tag{9.55}$$

We draw attention to the character of the solutions at small and large radii: as $r \to 0$, $u^2 \to \frac{2GM}{r}$, i.e. the gas is essentially in free fall, whereas at $r = \infty$, the condition that $u = 0 \Rightarrow \rho_s = \rho_\infty\, e^{1.5}$.

Hence for a given ρ_∞ we know ρ_s and thus for a given M and T we know $\dot{M}$. So if we place a star in an isothermal medium at rest and with a given ρ at ∞, we can calculate the accretion rate onto the star.

For example, consider a solar mass star located in an interstellar gas cloud of density 10^6 hydrogen atoms per cubic metre, and a temperature of 200 K. In this case, the sonic transition occurs at $r_s = GM/2c_s^2 \sim 4 \times 10^{13}$ m. Hence $\dot{M} = 4\pi\rho_s r_s^2 c_s \sim 10^{11}$ kg s^{-1}, or 3×10^{18} kg year^{-1}. This compares with a solar mass of 2×10^{30} kg. So from a cool cloud with such densities (comparable to the average density of gas in the Milky Way) the Sun would accrete about 1% of its mass on a timescale comparable with the age of the Universe (10^{10} years). In the dense cores of molecular clouds, gas densities are up to a million times higher so the accretion timescale drops to a short and astronomically interesting timescale. However, it also turns out that the *self-gravity* of the gas is important in these dense cloud cores, so that the formation of the Sun is usually thought of in terms of gravitational instability and collapse rather than Bondi accretion (see Chapter 10).

II Polytropic case

The Equation (9.53)

$$\frac{1}{2}u^2 + c_s^2 \ln\rho - \frac{GM}{r} = \frac{1}{2}c_s^2 + c_s^2 \ln\rho_s - \frac{GM}{r_s} \tag{9.56}$$

did not assume anything about the equation of state, so applies here as well. Again the sonic point is at

$$r_s = \frac{GM}{2c_s^2} \tag{9.57}$$

but now c_s has to be calculated because it is a function of ρ:

$$c_s^2 = \left(\frac{n+1}{n}\right) K\rho^{\frac{1}{n}}. \tag{9.58}$$

We know

$$\rho_s = \frac{\dot{M}}{4\pi r_s^2 c_s}, \tag{9.59}$$

and so

$$r_s = \frac{GM}{2c_s^2} = \left(\frac{\dot{M}}{4\pi\rho_s c_s^2}\right)^{\frac{1}{2}} \tag{9.60}$$

$\Rightarrow$

$$c_s^2 = \left(\frac{GM}{2}\right)^{\frac{4}{3}} \left(\frac{4\pi\rho_s}{\dot{M}}\right)^{\frac{2}{3}}. \tag{9.61}$$

Combining the two equations for c_s^2 (for $n \neq 2/3$) gives us

$$\rho_s = \left(\frac{GM}{2}\right)^{\frac{4n}{3-2n}} \left(\frac{4\pi}{\dot{M}}\right)^{\frac{2n}{3-2n}} \left(\frac{n}{(n+1)K}\right)^{\frac{3n}{3-2n}}. \tag{9.62}$$

So

$$c_s = \left(\frac{GM}{2}\right)^{\frac{2}{3-2n}} \left(\frac{4\pi}{\dot{M}}\right)^{\frac{1}{3-2n}} \left(\frac{n}{(n+1)K}\right)^{\frac{n}{3-2n}} \tag{9.63}$$

and

$$r_s = \frac{GM}{2\left[\frac{\pi(GM)^2}{\dot{M}}\right]^{\frac{2}{3-2n}} \left[\frac{n}{(n+1)K}\right]^{\frac{2n}{3-2n}}}. \tag{9.64}$$

Now, as before, we use Bernoulli's equation to determine how ρ and c_s behave as functions of r:

$$\frac{1}{2}u^2 + (n+1)K\rho^{\frac{1}{n}} - \frac{GM}{r} = \text{constant}, \tag{9.65}$$

where we can now evaluate the constant at r_s, since at that point $u = c_s$, $(n+1)K\rho^{\frac{1}{n}} = nc_s^2$ and $\frac{GM}{r} = 2c_s^2$. Therefore

$$\frac{1}{2}u^2 + (n+1)K\rho^{\frac{1}{n}} - \frac{GM}{r} = \left(n - \frac{3}{2}\right)c_s^2. \tag{9.66}$$

Substituting for u from the continuity equation $\Rightarrow$

$$\frac{1}{2}\left(\frac{\dot{M}}{4\pi r^2 \rho}\right)^2 + (n+1)K\rho^{\frac{1}{n}} = (n-\frac{3}{2})c_s^2 + \frac{GM}{r}. \tag{9.67}$$

If as $r \to \infty\, u \to 0$, then

$$\rho_\infty = \left[\frac{(n-\frac{3}{2})c_s^2}{(n+1)K}\right]^n \tag{9.68}$$

and

$$c_{s_\infty}^2 = \frac{(n-\frac{3}{2})}{n}c_s^2. \tag{9.69}$$

So, again, if we know ρ_∞ and the sound speed there, along with the mass of the central star, then we know the accretion rate $\dot{M}$:

$$\dot{M} = \frac{\pi(GM)^2\rho_\infty}{c_{s_\infty}^3}\left(\frac{n}{n-\frac{3}{2}}\right)^{n-\frac{3}{2}}. \tag{9.70}$$

Equation (9.66) implies that Bernoulli's constant is less than zero at r_s if $n < \frac{3}{2}$, but we know this constant is > 0 at $r = \infty$ if the density $\rho > 0$. This tells us that for $n < \frac{3}{2}$ the sonic point is never attained. The physical reason for this is that the gas is too incompressible – ∇p (directed outward) retards the flow enough to keep it subsonic everywhere, and it never reaches free fall.

A special case is $n = \frac{3}{2}$, and it is of interest because it is the one which corresponds to adiabatic flow in a monatomic gas ($\gamma = 1 + \frac{1}{n} = \frac{5}{3}$). As n is reduced towards the value of $\frac{3}{2}$ from above, Equation (9.69) implies that the sound speed at the sonic point becomes arbitrarily high and that (from Equation (9.52)) $r_s \to 0$. As we approach the condition $n = 3/2$, Equation (9.61) (which gives the ratio of squared sound speed to density to the two-thirds power, all evaluated at the sonic point) is still valid. However, for $n = 3/2$ the ratio of squared sound speed to density to the two-thirds power is just a constant, related to the adiabatic constant via (9.58). Therefore in the limit one can use (9.61) and (9.58) to relate K and $\dot{M}$, i.e. $K = \frac{3}{5}\left(\frac{GM}{2}\right)^{\frac{4}{3}}\left(\frac{4\pi}{\dot{M}}\right)^{\frac{2}{3}}$. Thus given conditions at infinity (i.e. density and temperature, which fix K) one may calculate $\dot{M}$ in this limiting case. (Note that in practice there is no problem with a singularity for this $n = 3/2$ case, because the accretor (star) always has finite size and so the sonic point (where the sound speed is formally infinite) is never reached in reality.)

We emphasise that nothing strange happens to the flow at the sonic point. It just marks a smooth transition between (pressure-dominated) subsonic and supersonic flow. There are no shocks, since they require sudden deceleration of supersonic flow (of course, one would expect a shock where the flow impinges the stellar surface, if this lies within r_{s}).

9.4 Stellar winds

Winds from the surfaces of stars are of great astrophysical interest for several reasons. First of all, the interactions of such winds with their environments, and the resulting shocks, can produce some of the most stunning images in astronomy, e.g. the Cat's Eye Nebula (Figure 9.7). Secondly, stellar winds are a significant source of *mechanical energy*

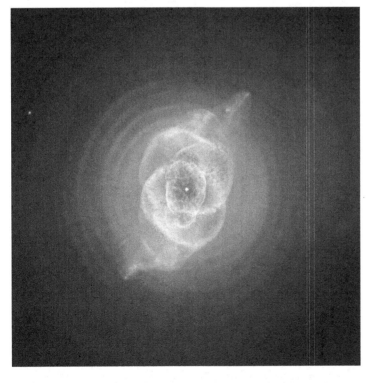

Fig. 9.7. The planetary nebula NGC 6543. Observations suggest the central star ejected its mass in a series of pulses at 1500-year intervals to make a layered, onion-skin structure extending out to $\sim 0.15\,\mathrm{pc}$ from the dying star. (NASA/STScI)

input into the ISM (for a wind with terminal velocity v_∞ and mass loss rate $\dot{M}_w$, the mechanical luminosity is $\frac{1}{2}\dot{M}_w v_\infty^2$). For example, in the case of a rich young star cluster, the input of mechanical energy into the ISM during the cluster's first 2 Myr of life is entirely from stellar winds from the most massive stars in the cluster, since no stars have evolved to the point of undergoing supernova explosions at such early times. The mechanical luminosity of the winds from these most massive stars is comparable to that which is delivered subsequently through successive supernova explosions in the cluster. The combined action of stellar winds and supernovae thus sustains a roughly constant mechanical luminosity for about 40 Myr (when the last supernova explodes), whose consequence is to blow large kiloparsec-scale cavities ('superbubbles') in the ISM. Figure 9.8 shows how the interstellar medium in the Large Magellanic Cloud is structured by the presence of such superbubbles. Since stars are generally clustered at birth, the creation of superbubbles through sustained mechanical energy input (winds + supernovae) probably has a larger effect on the ISM than the aggregate effect of isolated supernova explosions, as we considered in Chapter 7. Finally,

Fig. 9.8. A surface brightness map of the 21 cm H I emission from the Large Magellanic Cloud, showing the filamentary, bubbly, and flocculent structure of the interstellar medium. At the distance of the LMC 1° corresponds to 0.8 kpc. (Kim *et al.*, *Astrophysical Journal Supplement*, **148**, 473, 2003)

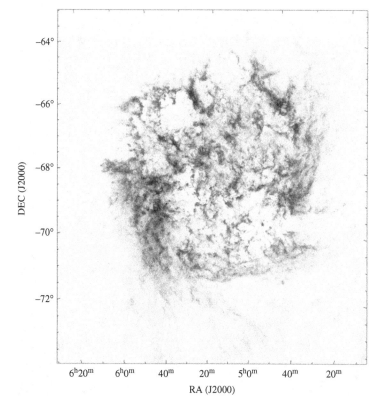

winds are an important channel for returning the products of stellar nucleosynthesis to the ISM. In the absence of winds, stars with masses less than $8 M_\odot$, which do not undergo supernovae, would be unable to recycle their gas into the ISM and the chemistry of the Universe would be very different. In particular, they provide a mechanism for returning *dust*, formed in the envelopes of cool stars, to the ISM. There is an obvious anthropic implication of this process, since the presence of dust in the ISM, and the possibility of its growth into rocky structures, is obviously indispensable to the existence of our own planet!

Winds are driven through the interaction between stellar radiation and momentum-absorbing species in the stellar atmosphere, although the microphysics of the interaction depends on the type of star concerned. For example, in OB stars, radiation is absorbed by metallic ions and thus the wind mass loss rate is a strong function of the metallicity. Wind mass loss rates from OB stars range over 10^{-6}–10^{-4} $M_\odot\,\mathrm{yr}^{-1}$, with terminal velocities which may attain thousands of km s^{-1}. In cooler stars, such as those on the Asymptotic Giant Branch (AGB), winds are instead driven by interaction between stellar radiation and dust grains that can form in the cool conditions of the wind flow. Wind mass loss rates in this case range over 10^{-8}–$10^{-4} M_\odot\,\mathrm{yr}^{-1}$, but with terminal velocities (10–45 km s^{-1}) that are much lower than in the case of OB star winds.

Mathematically, stellar winds are just the inverse problem to spherical accretion onto stars. The equations are the same – in the continuity equation $\dot{M}$ becomes the mass loss rate and u the outflow velocity. Bernoulli's equation contains only u^2 so is completely symmetric. The driving force (see above) results in extra terms in the momentum equation where the acceleration occurs, but if we look outside the acceleration region then we can consider the flow under pressure forces and gravity only, i.e. as above. The only difference is that the boundary conditions are set at the inner boundary, e.g. the density at the surface of the star. Although the steady state solutions developed here are useful tools for the analysis of stellar winds, we note that in practice such winds are highly variable, presumably as a result of time-dependent injection of material into the base of the wind flow. The possibility that material flowing at different speeds is able to catch up with material ahead of it results in internal shocks in the wind (in contrast with the shock-free solutions presented here: see similar remarks in the case of jets in Section 9.2). In the case of the Sun, it has been suggested that so-called coronal mass ejection events (in which material is intermittently ejected from the solar atmosphere into the interplanetary medium) are a consequence of variable injection of material into the base of the solar wind. In practice, it is therefore necessary to solve stellar wind flows as a time-dependent hydrodynamical problem.

9.5 General steady state solutions

When we were looking at the spherical flow solutions we did not need to impose the condition that $u \to 0$ as $r \to \infty$. For the isothermal case ($c_s = $ constant) the general solution to Equation (9.51)

$$\left(u^2 - c_s^2\right) \frac{\mathrm{d} \ln u}{\mathrm{d}r} = \frac{2c_s^2}{r} \left[1 - \frac{GM}{2c_s^2 r}\right] \tag{9.71}$$

is

$$\left(\frac{u}{c_s}\right)^2 - \ln\left(\frac{u}{c_s}\right)^2 = 4 \ln \frac{r}{r_s} + \frac{4r_s}{r} + C. \tag{9.72}$$

Note that, depending on the value of C, we do not necessarily have the flow making a sonic transition at $r = GM/2c_s^2$ (which was the solution we looked at in detail above) but can also have solutions in which the flow instead attains a maximum or minimum velocity at this radius. The nature of the possible solutions is sketched in Figure 9.9.

Fig. 9.9.

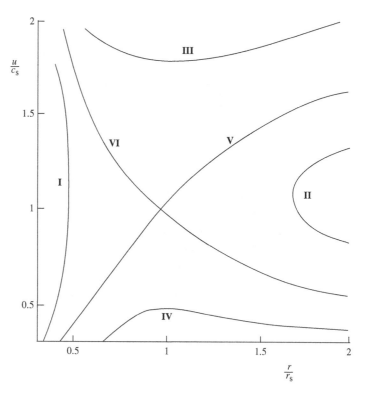

Some of these solutions are not physical – those curves in the regions marked **I** and **II** have two values of u at a given r. For solutions in region **III** $u > c_s$ always, so the flows are supersonic everywhere, and region **IV** solutions are subsonic everywhere. Only two curves have a transition between subsonic and supersonic flow, **V** and **VI**. For these $C = -3$, which one can readily see by noting that in Equation (9.71) this choice ensures $u = c_s$ and $r = GM/2c_s^2$. Which particular solution applies depends on the boundary conditions. Solution **V** applies for a subsonic wind flowing out from a star which accelerates to high speeds at large distances, and **VI** to the spherical accretion (Bondi accretion) starting from small speeds at ∞.

One point that often causes some misunderstanding about this sort of plot is what is the physical – as opposed to mathematical – significance of solutions in **I** and **II** – how does the flow behave in practice if it is given boundary conditions that place it in this region of the plot? Does the flow turn around or jump to another velocity when it finds it cannot get past the sonic point? We emphasise here that this problem is only encountered because we have *forced* the flow to obey steady state equations (i.e. we have discarded the $\frac{\partial}{\partial t}$ terms in order to derive Equation (9.71)). In practice, a flow set up with an initial value of C that places it in this part of the plot would not be steady and would self-adjust so as to yield steady state solutions lying in other regions of the plot.

Chapter 10
Fluid instabilities

Consider a fluid in a steady state, i.e. one which satisfies the hydro-dynamic equations with $\partial/\partial t = 0$ everywhere. If we find that small perturbations to this configuration grow with time, then our chances of finding the initial configuration in nature are very small, and the configuration is said to be unstable with respect to those perturbations. A stable configuration is one where either the perturbations diminish, or there is the possibility of oscillations or waves about the equilibrium configuration.

In this chapter we will be examining a variety of fluid instabil-ities. Such instabilities can often be invoked to explain the wealth of structure in astronomical images, as seen in e.g. Figures 9.7 and 9.8. More profoundly, the instabilities we discuss are responsible for such fundamental processes as convection in stars and the creation of the multi-phase state of the interstellar medium. Most importantly, the Jeans instability (Section 10.3) is responsible for the formation of the most important building blocks of the Universe – galaxies and stars.

10.1 Convective instability

The stability condition in the first case we consider is one which may be obtained by fairly simple arguments, and in particular without doing a perturbation analysis of the full hydrodynamic equations. It is also a case of considerable astrophysical importance, as we shall see.

Suppose we have a perfect gas in hydrostatic equilibrium in a uniform gravitational field. We choose the z axis so that gravity acts in the $-z$ direction, so the pressure $p(z)$ and the density $\rho(z)$ decrease as z increases. We take a fluid element at the same density and pressure as its surroundings, and displace it upward by a small amount δz,

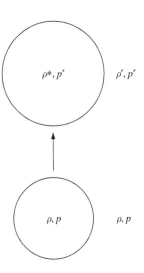

Fig. 10.1.

$\rho*, p'$ ρ', p'

ρ, p ρ, p

where the surrounding density and pressure are ρ' and p', as illustrated in Figure 10.1. We know that pressure imbalances are removed very quickly by acoustic waves, but that heat exchange takes considerably longer, so initially the region of gas will change adiabatically to be in pressure equilibrium at the new position. As a result, the region will have a new density $\rho*$ at the new position. Then, if $\rho* < \rho'$, the displaced region will be buoyant (according to Archimedes' principle) and will continue to move away from the initial position, so the system is unstable. If $\rho* > \rho'$ then the region will try to sink back to its original position, so the system is stable.

Since the region is displaced adiabatically, we have

$$\rho^* = \rho \left(\frac{p'}{p}\right)^{\frac{1}{\gamma}},$$

(10.1)

and if the pressure gradient is $\frac{dp}{dz}$ then

$$p' = p + \frac{dp}{dz}\delta z$$

(10.2)

to first order. Therefore

$$\rho^* = \rho + \frac{\rho}{\gamma p}\frac{dp}{dz}\delta z.$$

(10.3)

For the medium outside the displaced element

$$\rho' = \rho + \frac{d\rho}{dz}\delta z.$$

(10.4)

So the condition that $\rho^* < \rho'$ if the system is unstable becomes $\frac{\rho}{\gamma p} \frac{dp}{dz} < \frac{d\rho}{dz}$ in the medium. Another way of looking at this is to note that the density of the surroundings is larger, as you go up, than it would be for an adiabatic structure; therefore the quantity $\frac{p}{\rho^\gamma}$ must decline with z if the system is unstable.

Conversely, the system is stable if $\frac{\rho}{\gamma p} \frac{dp}{dz} > \frac{d\rho}{dz}$, or equivalently if $\frac{p}{\rho^\gamma}$ increases with z. It is neutrally stable if $\frac{p}{\rho^\gamma} = $ constant, i.e. $\frac{dK}{dz} = 0$, the isentropic case.

We can convert this stability criterion (the Schwarzschild criterion) to one linking the temperature and pressure gradients in the medium. Since $p = \frac{\mathcal{R}_*}{\mu} \rho T$, we have

$$\rho' = \rho + \left[\frac{\rho}{p} \frac{dp}{dz} - \frac{\rho}{T} \frac{dT}{dz} \right] \delta z. \tag{10.5}$$

So, using the expression for ρ^* above, we find

$$\rho^* - \rho' = \left[-\left(1 - \frac{1}{\gamma}\right) \frac{\rho}{p} \frac{dp}{dz} + \frac{\rho}{T} \frac{dT}{dz} \right] \delta z. \tag{10.6}$$

Then, since both $\frac{dp}{dz}$ and $\frac{dT}{dz}$ are negative, the medium is stable if

$$\left| \frac{dT}{dz} \right| < \left(1 - \frac{1}{\gamma}\right) \frac{T}{p} \left| \frac{dp}{dz} \right|. \tag{10.7}$$

This is the way the Schwarzschild criterion is normally expressed.

If the system is unstable, the upwardly displaced elements keep on going to form convection cells, whose size is set by the lengthscale over which elements cease to be adiabatic, i.e. exchange heat with their surroundings. The upwardly displaced elements are hotter than their surroundings (same pressure, lower density, so higher temperature) so heat exchange results in energy being deposited in the surrounding medium.

For downward displacements the result is similar, of course. If the system is unstable the regions keep on going, at lower temperatures than their surroundings, and are heated on some lengthscale. The net result is the formation of convection cells where convection transports heat upwards by carrying it in the displaced elements and then releasing it into the surroundings at the top, as illustrated in Figure 10.2.

Astrophysical applications

If a star is in complete hydrostatic equilibrium then all the heat transport is by radiation (as noted in Chapter 1, conduction is not normally important). We can ask whether or not the structure is stable to convection at each radius, and, if it is not, then the star is convective there.

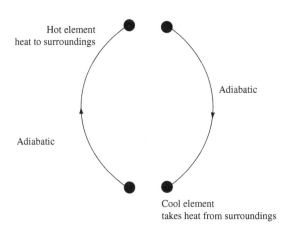

Fig. 10.2.

Hot element
heat to surroundings

Adiabatic

Adiabatic

Cool element
takes heat from surroundings

If this occurs, then to a good approximation the temperature gradient there may be replaced by the adiabatic temperature gradient. The justification for this statement involves determining the energy flux by convection, and comparing this with the radiative flux. In outline, the argument is that *if* the stellar structure were isentropic, then no energy would be transported by convection, since the displaced element would always be at the same temperature as its surroundings. It turns out that the energy carried by convection can be expressed in the form $k \times (\frac{dT}{dz}|_{ad} - \frac{dT}{dz})$, where k is some coefficient and the term in brackets is the difference between the actual value of the temperature gradient and the value it would have if the structure were isentropic. One calculates the value of k by making some assumption about how far a fluid element travels before coming back into thermal equilibrium with its surroundings, which in turn determines how much heat it deposits in its surroundings. (This 'mixing length' theory of convection is covered in any book on stellar structure; see for example Collins, *The Fundamentals of Stellar Astrophysics*, W. H. Freeman & Co., 1989.) For our present purposes, it is enough to note that k turns out to be so large that the required heat flux can be carried when the difference between the actual temperature gradient and the isentropic one is exceedingly small. In this limit, it is acceptable to set the two gradients equal to each other, as far as evaluating the *structure* of the star is concerned, although remaining mindful of the fact that, as far as *thermal* transport in the star is concerned, it is the small deviation between the two temperature gradients that allows heat to be transported by convective motions.

Under what circumstances do real stars develop convective regions? The stability criterion (Equation (10.7) above) shows us that convection is associated with regions of the stars where the temperature declines steeply with increasing radius. (One can see why this is, heuristically, by seeing that a hydrostatic structure with dramatically decreasing temperature as one goes out will have a structure in which the density does not decline very steeply, since otherwise gravity would not be able to balance the large outwardly directed pressure gradient. If, however, the density of the unperturbed structure is not decreasing steeply as one goes out, the perturbed element is more likely to be buoyant in relation to that background.) An example of a situation where the temperature would decline steeply in the absence of convection is in the cores of massive stars, where this steep temperature decline is associated with the very strong positive temperature dependence of nuclear energy generation via the CNO process. Another example is in the envelope of low mass stars which are cool enough to contain regions of partial ionisation. Partially ionised regions are associated with large opacities and a given radiative flux can only be driven through these opaque regions if the temperature gradient is very steep there. Consequently, the outer regions of low mass stars and the cores of high mass stars are convective, whereas intermediate mass stars (of around $3\,M_\odot$) have entirely radiative structures. These issues are amplified considerably in any book on stellar structure.

Stable configuration

The force per unit volume acting on the displaced material is $-g(\rho^* - \rho')$, so the equation of motion for a fluid element can be written as

$$\rho^* \frac{\mathrm{d}^2}{\mathrm{d}t^2} \delta z = -g(\rho^* - \rho'), \tag{10.8}$$

providing it does not disturb the surrounding gas (hardly likely!). We have

$$\rho^* - \rho' = \left[-\left(1 - \frac{1}{\gamma}\right) \frac{\rho}{p} \frac{\mathrm{d}p}{\mathrm{d}z} + \frac{\rho}{T} \frac{\mathrm{d}T}{\mathrm{d}z} \right] \delta z, \tag{10.9}$$

so this becomes, for small displacements,

$$\frac{\mathrm{d}^2}{\mathrm{d}t^2} \delta z + \frac{g}{T} \left[\frac{\mathrm{d}T}{\mathrm{d}z} - \left(1 - \frac{1}{\gamma}\right) \frac{T}{p} \frac{\mathrm{d}p}{\mathrm{d}z} \right] \delta z = 0. \tag{10.10}$$

If the stability condition is satisfied then

$$N = \sqrt{\frac{g}{T} \left[\frac{\mathrm{d}T}{\mathrm{d}z} - \left(1 - \frac{1}{\gamma}\right) \frac{T}{p} \frac{\mathrm{d}p}{\mathrm{d}z} \right]} \tag{10.11}$$

is a frequency associated with the oscillations about the equilibrium configuration. So in this approximation we have a frequency for internal gravity waves in a stratified atmosphere. (They are called internal gravity waves to distinguish them from surface gravity waves which can take place on surfaces of liquids in air.)

10.2 Rayleigh-Taylor and Kelvin-Helmholtz instabilities

These are instabilities associated with layers of fluid with a discontinuous change in tangential velocity and/or density across an interface. This class of instabilities accounts, for example, for the over-turning of fluid in the case that dense material over-lies lighter material in the presence of gravity and also accounts for the buckling of interfaces subject to shear motions. Such phenomena are observed in many settings, including astrophysics, atmospheric and ocean science and industrial processes.

Consider two fluids one lying above the other at rest in a uniform gravitational field, with the lower one having density ρ just below the interface, and the upper one density ρ' just above. We consider how perturbations at the interface develop, and restrict ourselves to two dimensions, one in the plane of the interface x and the other perpendicular to it z. For simplicity, we will consider a flow which is incompressible and irrotational, and as a consequence we can describe the velocity field by a potential function where $u = -\nabla\Phi$. Under these circumstances the Eulerian momentum equation becomes

$$-\nabla\frac{\partial\Phi}{\partial t} + \nabla\left(\frac{1}{2}u^2\right) = -\nabla\left(\frac{p}{\rho}\right) - \nabla\Psi, \qquad (10.12)$$

where Ψ is the gravitational potential as usual. This can be integrated to give

$$-\frac{\partial\Phi}{\partial t} + \left(\frac{1}{2}u^2\right) + \left(\frac{p}{\rho}\right) + \Psi = F(t), \qquad (10.13)$$

where $F(t)$ is a constant in space but can be a function of time.

The configuration of interest is shown in Figure 10.3. The horizontal plane $z = 0$ separates the two fluids, and we assume they have uniform velocities U and U' in the x direction. This configuration satisfies the steady state hydrodynamic equations, as may be easily

Fig. 10.3.

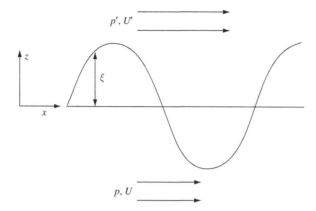

verified. We consider perturbations in the surface of the interface so the perturbed position is $\xi(x, t)$. We aim to find out if the perturbation grows, oscillates or decays with time.

The velocity potential in the fluid below can be written as

$$\Phi = -Ux + \phi, \tag{10.14}$$

where ϕ is the perturbed part, which satisfies

$$\nabla^2\phi = 0 \tag{10.15}$$

as a result of the incompressibility condition $\nabla \cdot \mathbf{u} = 0$.

Similarly, in the upper fluid

$$\Phi' = -U'x + \phi', \tag{10.16}$$

with

$$\nabla^2\phi' = 0. \tag{10.17}$$

The velocity perturbations are caused by displacements of the interface, so we now need to connect the velocity potential perturbations ϕ and ϕ' with ξ. If we take a fluid element within the lower fluid and at the interface, then its vertical velocity is given by $-\partial\phi/\partial z$. The velocity of this particular element is also given by the Lagrangian derivative of the displacement, $D\xi/Dt$. Therefore, converting the Lagrangian derivative to an Eulerian one in the usual way, we have

$$-\frac{\partial\phi}{\partial z} = \frac{\partial\xi}{\partial t} + U\frac{\partial\xi}{\partial x} \tag{10.18}$$

at $z = 0$, to first order in the perturbed quantities. In exactly the same way we have, for the fluid just above the interface,

$$-\frac{\partial \phi'}{\partial z} = \frac{\partial \xi}{\partial t} + U' \frac{\partial \xi}{\partial x} \tag{10.19}$$

at $z = 0$. Because we are linearising the perturbation equations, any arbitrary perturbation may be written as the sum of Fourier components, so as with sound waves we seek solutions of the form

$$\xi = A \mathrm{e}^{\mathrm{i}(kx - \omega t)}, \tag{10.20}$$

and ϕ and ϕ' will have the same x and t dependences. For both ϕ and ϕ' Laplace's equation has to be satisfied, from above, and this sets the z dependence so

$$\phi = C\, \mathrm{e}^{\mathrm{i}(kx - \omega t) + kz}, \tag{10.21}$$

and

$$\phi' = C'\, \mathrm{e}^{\mathrm{i}(kx - \omega t) - kz}, \tag{10.22}$$

where the signs before the kz terms have been chosen so that the perturbations do not grow exponentially as we go far away from the interface.

Now substitute these expressions in the equations linking ξ with the ϕs, (10.18) and (10.19), to get

$$\mathrm{i}(kU - \omega)A = -kC, \tag{10.23}$$

$$\mathrm{i}(kU' - \omega)A = kC'. \tag{10.24}$$

So we have two equations for three unknowns. However, we have not used the condition that the pressure must be continuous across the interface. We have an expression involving the pressure for incompressible irrotational flows above (Equation (10.13))

$$-\frac{\partial \Phi}{\partial t} + \left(\frac{1}{2}u^2\right) + \left(\frac{p}{\rho}\right) + \Psi = F(t), \tag{10.25}$$

so the pressure inside the lower fluid at the interface is

$$p = -\rho\left(-\frac{\partial \phi}{\partial t} + \frac{u^2}{2} + g\xi\right) + \rho F(t), \tag{10.26}$$

where the gravitational potential has been written as $g\xi$. Then, using a similar expression for the fluid above the interface, we have

$$\rho\left(-\frac{\partial\phi}{\partial t}+\frac{u^2}{2}+g\xi\right)=\rho'\left(-\frac{\partial\phi'}{\partial t}+\frac{u'^2}{2}+g\xi\right)+K \tag{10.27}$$

at $z=0$, where $K=\rho F(t)-\rho' F'(t)$.

In principle K is a function of time, but here it is a constant because of the boundary condition that the perturbations vanish as $z\to\pm\infty$ at all times (using (10.13)). Then we can use the unperturbed values $(u=U, u'=U', \phi=\phi'=\xi=0)$ to determine K, which is then

$$K=\frac{1}{2}\rho U^2-\frac{1}{2}\rho' U'^2. \tag{10.28}$$

We still have to determine u^2 and u'^2 for Equation (10.27), but these are determinable directly from (10.14), so

$$u^2=(U\hat{\mathbf{e}}_x-\nabla\phi)^2=U^2-2U\frac{\partial\phi}{\partial x} \tag{10.29}$$

to first order. A similar expression also holds for u', so these with (10.18) and (10.19) give

$$\rho\left(-\frac{\partial\phi}{\partial t}-U\frac{\partial\phi}{\partial x}+g\xi\right)=\rho'\left(-\frac{\partial\phi'}{\partial t}-U'\frac{\partial\phi'}{\partial x}+g\xi\right) \tag{10.30}$$

at $z=0$.

Now we can substitute from Equations (10.21) and (10.22) in this, to obtain the required third equation connecting the amplitudes of the perturbed quantities:

$$\rho[-i(kU-\omega)C+gA]=\rho'[-i(kU'-\omega)C'+gA]. \tag{10.31}$$

Therefore, with (10.23) and (10.24), we find the dispersion relation

$$\rho(kU-\omega)^2+\rho'(kU'-\omega)^2=kg(\rho-\rho'). \tag{10.32}$$

If we are given k or ω, then the other variable is given by the solution to the quadratic equation. For a given k, we find the phase velocity of the wave is given by

$$\frac{\omega}{k}=\frac{\rho U+\rho' U'}{\rho+\rho'}\pm\left[\frac{g}{k}\frac{\rho-\rho'}{\rho+\rho'}-\frac{\rho\rho'(U-U')^2}{(\rho+\rho')^2)}\right]^{\frac{1}{2}}. \tag{10.33}$$

This result is a general one which we apply to some particular cases.

Surface gravity waves

Take two fluids at rest, with $\rho' < \rho$, and then

$$\frac{\omega}{k} = \pm \sqrt{\frac{g}{k} \frac{\rho - \rho'}{\rho + \rho'}}. \tag{10.34}$$

Then ω is real for real k, and the disturbance moves as a wave – these are called surface gravity waves. The velocity $\frac{\omega}{k}$ depends on the wavenumber k, so different wavelengths propagate at different speeds – the waves are dispersive.

If $\rho' \ll \rho$ (e.g. air and water), then $\omega = \pm\sqrt{gk}$, or the phase velocity $\frac{w}{k} \propto 1/\sqrt{k}$.

Static stratified fluid under gravity

Suppose now the density in the fluid above is greater than that in the fluid below, i.e. $\rho' > \rho$. The hydrostatic equation can be satisfied under these circumstances, but we all know that this situation is unstable.

Under these circumstances,

$$\omega = \pm \, ik \sqrt{\frac{g}{k} \frac{\rho' - \rho}{\rho + \rho'}} \tag{10.35}$$

is imaginary for real k, so in the solution

$$\xi = A \, e^{i(kx - \omega t)} \tag{10.36}$$

the positive imaginary solution of ω results in an exponentially growing mode. This is the true Rayleigh–Taylor instability.

Since the uniform acceleration in a mechanical system is equivalent to a gravitational field, the problem of a light fluid accelerating into a heavy fluid is very similar and gives rise to the same instability.

Astrophysical application

A (qualitative) application is to a supernova explosion. We saw in the blast wave case that we have a thin shell of gas decelerating outwards. Inward acceleration is equivalent to outward directed gravity in the rest frame of the blast wave interface, so we have the dense shell of gas 'on top' of the less dense gas outside. In this case too there is the equivalent of a Rayleigh–Taylor instability, leading to the production of filaments in the post-shock gas.

This instability is relevant earlier in the supernova phase as well. As the star explodes a decelerating shock wave passes outward through the star, and this is similarly equivalent to reversing the direction of the effective gravity in what had been a stably stratified star. As a consequence the star becomes Rayleigh–Taylor unstable, and the envelope is thoroughly mixed by this instability.

A yet further example of Rayleigh–Taylor instabilities during the evolution of supernova remnants occurs at the stage that the gas swept up in the shock first begins to cool significantly and collapses into a thin shell. At that point, material is *accelerated outwards* towards the developing shell by the outwardly directed pressure gradient. This outward acceleration is equivalent to inwardly directed gravity, and yet the shell material is denser than the gas interior to it. Consequently, Rayleigh–Taylor instabilities occur at this stage also.

Kelvin–Helmholtz instability

We consider the case when U and U' are not zero, and suppose that $\rho > \rho'$ so the system is Rayleigh–Taylor stable. If the expression within the square root of Equation (10.33) is negative, then for real k it follows that ω has an imaginary part. We know a positive imaginary part leads to instability, and this instability will arise when the term within the square root is negative, i.e. when

$$\frac{g}{k}\frac{\rho - \rho'}{\rho + \rho'} - \frac{\rho\rho'(U - U')^2}{(\rho + \rho')^2)} < 0, \tag{10.37}$$

i.e.

$$\rho\rho'(U - U')^2 > (\rho^2 - \rho'^2)\frac{g}{k}. \tag{10.38}$$

This instability is called the Kelvin–Helmholtz instability, and it causes the interface between two fluids to wrinkle if they are travelling at different speeds.

Note that the instability occurs for

$$k > \frac{(\rho^2 - \rho'^2)g}{\rho\rho'(U - U')^2}, \tag{10.39}$$

so if k is large enough (i.e. the wavelength is small enough) for any finite velocity difference then the perturbations are unstable. (Short wavelength perturbations can however be damped by surface tension, so for fluids we are used to there is a critical value of the velocity difference below which the flow is stable.)

It follows from the inequality (10.38) that if $g = 0$ then *any* wavenumber is Kelvin–Helmholtz unstable. An extragalactic or stellar jet moving at high speed with respect to the surrounding medium would therefore be subject to Kelvin–Helmholtz instability. Many jets do show variations in intensity and cross-section along the axis of the jet, and a possible interpretation is that these are a consequence of this instability.

10.3 Gravitational instability (Jeans instability)

In the derivation where the propagation of sound waves was examined we neglected gravity, except as a constant external force when we dealt with the propagation of sound waves in an isothermal atmosphere. Now we consider the possibility of self-gravity playing a role, so perform the same analysis as for the sound wave in a uniform medium (in the absence of external gravity), but now involve the extra variable $\Delta\Psi$ (the perturbed gravitational potential) and so an extra equation linking $\Delta\Psi$ to the density perturbation $\Delta\rho$ (Poisson's equation). We therefore consider a sound wave passing through a uniform medium (constant p, ρ) with wavelength long enough that we can no longer ignore gravity.

We have, as usual,

$$\frac{\partial\rho}{\partial t} + \nabla\cdot(\rho\mathbf{u}) = 0, \tag{10.40}$$

$$\frac{\partial\mathbf{u}}{\partial t} + \mathbf{u}\cdot\nabla\mathbf{u} = -\frac{1}{\rho}\nabla p - \nabla\Psi, \tag{10.41}$$

and

$$\nabla^2\Psi = 4\pi G\rho. \tag{10.42}$$

As before, we take $p = p_0 + \Delta p$, $\rho = \rho_0 + \Delta\rho$, $u = \Delta u$ and now $\Psi = \Psi_0 + \Delta\Psi$. Then, using the Eulerian equations above, and retaining only terms to first order in Δanything:

$$\frac{\partial\Delta\rho}{\partial t} + \rho_0\nabla\cdot\Delta\mathbf{u} = 0, \tag{10.43}$$

$$\frac{\partial\Delta\mathbf{u}}{\partial t} = -\frac{dp}{d\rho}\frac{\nabla\Delta\rho}{\rho_0} - \nabla\Delta\Psi = -c_s^2\frac{\nabla\Delta\rho}{\rho_0} - \nabla\Delta\Psi, \tag{10.44}$$

and

$$\nabla^2\Delta\Psi = 4\pi G\Delta\rho. \tag{10.45}$$

As before, the equations in the unperturbed quantities have to be satisfied, so those terms disappear.

Actually, though, the assumption of a uniform static medium is not consistent with the equations. We need

$$\nabla p_0 = -\rho_0\nabla\Psi_0 \tag{10.46}$$

and

$$\nabla^2\Psi_0 = 4\pi G\rho_0. \tag{10.47}$$

If p_0 is a constant then Ψ_0 must also be constant from the first of these, and if Ψ_0 is constant then the second of the two equations implies $\rho_0 = 0$. This is equivalent to the statement that a static universe is empty. In order to do the job properly, one should first find a self-consistent potential and density distribution for the initial hydrostatic configuration and then consider perturbations about that solution. The enthusiastic could, for example, consider an isothermal self-gravitating slab as the initial hydrostatic equilibrium state (see Chapter 5), but here we will adopt the approach adopted by Jeans in 1902 and pretend that a uniform density constant gravitational potential cloud satisfies the equilibrium equations. This approach is sometimes referred to as the *Jeans swindle*! It has the merit that it is simple, and gives results which are qualitatively similar to true configurations of interest. The reason that we do not care very much is that the outcome of our analysis will be a scale where self-gravity in a sound wave is important, and getting it wrong by factors of order unity does not really matter.

So, we use Equations (10.43)–(10.45), and, as before, we write

$$\Delta\rho = \rho_1 e^{i(\mathbf{k}\cdot\mathbf{x}-\omega t)}, \tag{10.48}$$

$$\Delta\Psi = \Psi_1 e^{i(\mathbf{k}\cdot\mathbf{x}-\omega t)}, \tag{10.49}$$

and

$$\Delta u = u_1 e^{i(\mathbf{k}\cdot\mathbf{x}-\omega t)}. \tag{10.50}$$

Then

$$-\rho_1\omega + \rho_0\mathbf{k}\cdot\mathbf{u}_1 = 0, \tag{10.51}$$

$$-\rho_0\omega\mathbf{u}_1 = -c_s^2\rho_1\mathbf{k} - \rho_0\Psi_1\mathbf{k}, \tag{10.52}$$

and

$$-k^2\Psi_1 = 4\pi G\rho_1. \tag{10.53}$$

Eliminating $\mathbf{u}_1$ using (10.51) and (10.52) gives

$$\rho_1\omega^2 = k^2(\rho_0 c_s^2 + \rho_0\Psi_1), \tag{10.54}$$

which, with (10.52), gives

$$\omega^2 = c_s^2\left(k^2 - \frac{4\pi G\rho_0}{c_s^2}\right) \tag{10.55}$$

or

$$\omega^2 = c_s^2 \left(k^2 - k_J^2 \right), \tag{10.56}$$

where

$$k_J^2 = \frac{4\pi G \rho_0}{c_s^2}. \tag{10.57}$$

Here ω is imaginary for real k, and so the system is unstable, when $4\pi G \rho_0 > k^2 c_s^2$. Pressure tends to damp out density fluctuations, but gravity compresses the matter further, and what we have here is a criterion for when gravity will dominate.

Thus there is a maximum stable wavelength, called the *Jeans length*,

$$\lambda_{\text{ms}} = \left(\frac{\pi c_s^2}{G \rho_0} \right)^{\frac{1}{2}}. \tag{10.58}$$

The total mass contained within this wavelength is $\sim \rho_0 \lambda_{\text{ms}}^3$ and is called the *Jeans mass*.

$$M_J \sim \frac{\pi^{\frac{3}{2}} c_s^3}{G^{\frac{3}{2}} \rho_0^{\frac{1}{2}}}. \tag{10.59}$$

In general the wave will be isothermal (it is a long wavelength slowly evolving perturbation generally), so the sound speed $c_s^2 = \frac{\mathcal{R}_*}{\mu} T$ and (10.59) becomes

$$M_J \sim \left(\frac{\pi \frac{\mathcal{R}_*}{\mu} T}{G} \right)^{\frac{3}{2}} \frac{1}{\rho_0^{\frac{1}{2}}}. \tag{10.60}$$

Another way of looking at the Jeans length is that it is the length-scale over which the sound-crossing time is the same as the free-fall time under self-gravity. For a length ℓ the free-fall time is $\sim \ell/v$, where $v \sim \sqrt{GM/\ell}$, and the mass $M \sim \rho_0 \ell^3$, so $t_{\text{ff}} \sim 1/\sqrt{G\rho_0}$. The sound-crossing time $t_s \sim \ell/c_s$, so equating these two gives us $\ell \sim c_s/\sqrt{G\rho_0}$, i.e. of order the Jeans length.

The Jeans instability is of fundamental importance in astrophysics – indeed, it is the basic reason why the Universe is not uniform. The wealth of self-gravitating structures in the Universe, from stars to galaxies, are all likely to have arisen from local over-densities of mass exceeding the local Jeans mass. As such regions began to collapse, sound waves would (by the above arguments) not have time to run ahead of the collapse and set up the pressure gradients required to off-set further collapse. Consequently, the instability becomes non-linear (i.e. the perturbed quantities are no longer small compared with the unperturbed quantities) and the above treatment breaks down. Since the advent of millimetre wave astronomy it has become straightforward in recent years to probe the internal structure of Giant Molecular Clouds and to identify dense *cores* of gas with temperatures and densities ($T \sim 10\,\mathrm{K}$, $n \sim 10^{11}\,\mathrm{m}^{-3}$) for which the Jeans mass is similar to the core mass which in turn is roughly a solar mass. Thus it would seem that we have in this case identified the Jeans unstable regions which are destined to become future stars. On the other hand, it is not possible to identify regions of the Universe that are destined to become tomorrow's galaxies, since conditions in the intergalactic medium at current epochs imply Jeans masses that far exceed galactic masses, implying that galaxy-scale objects must have condensed when the Universe was considerably denser than it is today. As far as another obvious class of self-gravitating structure in the Universe – planets – is concerned, opinion is currently divided as to whether the Jeans instability has played any role in their creation. A popular model for the creation of gas giant planets (such as Jupiter) is that their initial growth was enabled by sticky collisions between icy grains in the primordial solar nebula, although other models invoke a Jeans-type instability in the nebula to achieve the initial collapse of the proto-planet.

10.4 Thermal instability

Thermal instability implies – as the name suggests – an instability leading to runaway heating or cooling following the perturbation of the temperature from an initial thermal equilibrium state. Evidently, whether a system is thermally unstable or not will depend on the physical processes that cool and heat the fluid and we will first examine these in more detail.

Which form you use depends on the circumstances. Here we will adopt the parameterised cooling law

$$\dot{Q} = A\rho T^{\alpha} - \mathcal{H}, \qquad (10.61)$$

which was discussed in Chapter 4. First, however, we set up the instability criterion for a general cooling law.

10.4.1 Thermal instability in astrophysical fluids

We start off with a gas in thermal equilibrium, so $\dot{Q} = 0$. A perturbation about thermal equilibrium could be characterised by a small temperature increase ΔT. If the gas heats up locally at constant pressure, then

$$\dot{Q} \rightarrow \dot{Q} + \frac{\partial \dot{Q}}{\partial T}\bigg|_p \Delta T. \tag{10.62}$$

If $\frac{\partial \dot{Q}}{\partial T}\big|_p < 0$ then the temperature perturbation grows $\Rightarrow$ *instability*.

While a constant pressure model is the most likely, since pressure adjustments usually take place quickly, we do not have to impose this restriction. We can do a full perturbation analysis as in any problem. As usual:

$$\frac{\partial \rho}{\partial t} + \nabla \cdot (\rho \mathbf{u}) = 0, \tag{10.63}$$

$$\rho \frac{\partial \mathbf{u}}{\partial t} + \rho \mathbf{u} \cdot \nabla \mathbf{u} = -\nabla p \tag{10.64}$$

(no gravity), and some suitable energy equation, like

$$T \frac{dS}{dt} = -\dot{Q}. \tag{10.65}$$

If we substitute for S, T in terms of K, ρ, then

$$\frac{dK}{dt} = -\frac{\gamma - 1}{\rho^{\gamma - 1}} \dot{Q}. \tag{10.66}$$

(This is not really an obvious result. We have

$$\begin{aligned} T\, dS &= C_V\, dT - \frac{\mathcal{R}_* T}{\mu} \frac{d\rho}{\rho} \\ &= \frac{\mathcal{R}_* T}{\mu} \left(\frac{1}{\gamma - 1} \frac{dT}{T} - \frac{d\rho}{\rho} \right). \end{aligned} \tag{10.67}$$

We also have

$$p = K\rho^\gamma = \frac{\mathcal{R}_*}{\mu} \rho T, \tag{10.68}$$

so

$$\frac{dp}{p} = \frac{d\rho}{\rho} + \frac{dT}{T} = \gamma \frac{d\rho}{\rho} + \frac{dK}{K}, \tag{10.69}$$

or

$$\frac{dK}{K} = -(\gamma - 1)\frac{d\rho}{\rho} + \frac{dT}{T}. \tag{10.70}$$

Comparison with Equation (10.67) shows that one may write

$$T\,dS = \frac{\mathcal{R}_* T}{\mu(\gamma - 1)} \frac{dK}{K}$$

$$= \frac{1}{\gamma - 1} \frac{p}{\rho} \frac{dK}{K}. \tag{10.71}$$

From the definition of $K\,(=\frac{p}{\rho^\gamma})$, we may re-write this as

$$T\,dS = \frac{\rho^{\gamma - 1}}{\gamma - 1}\,dK \tag{10.72}$$

and hence we recover Equation (10.66).)

Consider first order perturbation about a static equilibrium in thermal balance, so $\mathbf{u}_0 = 0$, $\dot{Q}_0 = 0$, $\nabla p_0 = 0$, $\nabla K_0 = 0$ and $\nabla \rho_0 = 0$, and consider the equations when the Lagrangian perturbations from these values are $\Delta\mathbf{u}$, $\Delta\rho$, etc. Since the unperturbed medium is uniform in all variables, the Eulerian perturbations (which must be substituted into the Eulerian fluid equations (10.63), (10.64), see Chapter 6) are simply equal to the Lagrangian perturbations.

Then, from (10.63),

$$\frac{\partial \Delta\rho}{\partial t} + \rho_0 \nabla \cdot \Delta\mathbf{u} = 0. \tag{10.73}$$

From (10.64)

$$\rho_0 \frac{\partial \Delta\mathbf{u}}{\partial t} = -\nabla \Delta p. \tag{10.74}$$

From (10.65)

$$\frac{\partial \Delta K}{\partial t} = -\frac{\gamma - 1}{\rho_0^{\gamma - 1}} \Delta\dot{Q}. \tag{10.75}$$

But we know

$$\Delta\dot{Q} = \left.\frac{\partial\dot{Q}}{\partial p}\right|_\rho \Delta p + \left.\frac{\partial\dot{Q}}{\partial\rho}\right|_p \Delta\rho, \tag{10.76}$$

so we can write the thermal equation (10.75) as

$$\frac{\partial \Delta K}{\partial t} = -A^* \Delta p - B^* \Delta\rho, \tag{10.77}$$

where

$$A^* = \frac{\gamma - 1}{\rho_0^{\gamma - 1}} \left.\frac{\partial\dot{Q}}{\partial p}\right|_\rho \tag{10.78}$$

and

$$B^* = \frac{\gamma - 1}{\rho_0^{\gamma-1}} \frac{\partial \dot{Q}}{\partial \rho}\bigg|_p. \tag{10.79}$$

Finally, we relate Δp to $\Delta \rho$ and ΔK, where we know

$$\frac{\Delta p}{p_0} = \frac{\Delta K}{K_0} + \gamma \frac{\Delta \rho}{\rho_0}. \tag{10.80}$$

Therefore

$$\Delta p = \rho_0^{\gamma} \Delta K + \gamma K \rho_0^{\gamma-1} \Delta \rho, \tag{10.81}$$

$$\Delta p = \rho_0^{\gamma} \Delta K + c_s^2 \Delta \rho, \tag{10.82}$$

where we have simply here *defined* c_s as the quantity relating adiabatic density and pressure perturbations, without, of course, assuming that the perturbations are adiabatic in this case (hence the term involving ΔK in (10.82) above).

Equations (10.73), (10.74), (10.77) and (10.82) are linear equations for the evolution of $\Delta \rho$, $\Delta \mathbf{u}$, ΔK, and Δp with constant coefficients. Since we are interested in growing modes we might as well replace the $-i\omega t$ we used when we sought oscillatory solutions by qt, and then if q is real and positive the system will be unstable. We may therefore write

$$\Delta \rho = \rho_1 e^{i\mathbf{k} \cdot \mathbf{x} + qt}, \tag{10.83}$$

$$\Delta \mathbf{u} = \mathbf{u}_1 e^{i\mathbf{k} \cdot \mathbf{x} + qt}, \tag{10.84}$$

etc. and substitute into these equations to obtain

$$q\rho_1 + \rho_0 i\mathbf{k} \cdot \mathbf{u}_1 = 0, \tag{10.85}$$

$$q\rho_0 \mathbf{u}_1 = -i\mathbf{k} p_1, \tag{10.86}$$

$$qK_1 = -A^* p_1 - B^* \rho_1, \tag{10.87}$$

$$p_1 = K_1 \rho_0^{\gamma} + c_s^2 \rho_1. \tag{10.88}$$

Eliminating $\mathbf{u}_1$ from (10.85) and (10.86), we get

$$q^2 \rho_1 = -k^2 p_1 \tag{10.89}$$

so (10.87) $\Rightarrow$

$$qK_1 = \frac{A^* q^2}{k^2} \rho_1 - B^* \rho_1 \tag{10.90}$$

and (10.88) $\Rightarrow$

$$\left(-\frac{q^2}{k^2} - c_s^2\right)\rho_1 = \rho_0^\gamma K_1.$$
(10.91)

So

$$\frac{A^*q}{k^2} - \frac{B^*}{q} = -\left(\frac{q^2}{k^2} + c_s^2\right)\frac{1}{\rho_0^\gamma}.$$
(10.92)

This has eliminated all the perturbed quantities, and left us with a relationship between k and q as we had with k and ω in previous examples – it is just a dispersion relation.

It is useful to rearrange this as a polynomial in q, i.e.

$$E(q) \equiv q^3 + A^*q^2\rho_0^\gamma + k^2c_s^2q - B^*k^2\rho_0^\gamma = 0.$$
(10.93)

There is an instability if there is a real positive root to the above equation, i.e. if the LHS is zero for $0 \le q \le \infty$.

Now $E(\infty) = \infty$, and $E(0) = -B^*k^2\rho_0^\gamma$, so the system is unstable if $B^* > 0$, i.e. if $\frac{\partial \dot{Q}}{\partial \rho}\big|_p > 0$. We can see that this is equivalent to the condition $\frac{\partial \dot{Q}}{\partial T}\big|_p < 0$ since

$$\frac{\partial \dot{Q}}{\partial \rho}\bigg|_p = \frac{\partial \dot{Q}}{\partial \left(\frac{\mu p}{\mathcal{R}_* T}\right)}\bigg|_p = -\frac{\mathcal{R}_* T^2}{\mu p}\frac{\partial \dot{Q}}{\partial T}\bigg|_p.$$
(10.94)

So the instability criterion is

$$\boxed{\frac{\partial \dot{Q}}{\partial T}\bigg|_p < 0. \qquad \textit{Field criterion} \qquad (10.95)}$$

Note that this is exactly the same result as the one we obtained before, when we made the restriction that the perturbation had to remain in pressure balance with the surroundings! The above derivation makes no such restriction, but allowed the perturbation to induce pressure and velocity changes coupled by the linearised fluid equations. It might appear surprising that the stability criterion expressed above only depends on the temperature derivative of the net cooling rate at constant pressure (and not on the other variable A^*, which is related to the temperature derivative of the net cooling rate at fixed *density*). Further examination of (10.93) shows us that actually the situation is indeed a little more complicated than indicated by the Field criterion. If the system is unstable according to the Field criterion, then it is indeed always unstable, regardless of the sign of the temperature derivative of the net cooling rate at fixed density. However, if it is stable according to the Field criterion, it *may* still be unstable if the temperature derivative of the net cooling rate at constant density is negative. Inspection of (10.93) shows that this destabilisation will be effective at low k, i.e. for long wavelength perturbations. In fact it turns out that a rough measure of the wavelength above which the A^* term can destabilise a system that is stable according to the Field criterion is provided by the distance travelled by a sound wave over the thermal timescale (the latter being the thermal energy divided by $\dot{Q}$). This makes sense, since in the case of such long wavelength perturbations, there is no time for sound waves to re-establish pressure balance with the surroundings over the thermal timescale on which the instability develops and hence the behaviour of the cooling rate at constant density is the relevant one. In practice, however, this situation (net cooling increasing with temperature at constant pressure but decreasing with temperature at constant density) is not usually encountered (since it requires that the cooling rate declines steeply with density). Therefore in practice the Field criterion provides a good determinant of thermal stability.

We can also use (10.93) to estimate the growth rate of the instability and how this depends on the wavelength of the perturbation. We consider now the case where the cooling at constant pressure is destabilising ($B^* > 0$) but the cooling at constant density is stabilising ($A^* > 0$). In the limit of long wavelength perturbations (small k) the root of (10.93) is determined by balance between the second and fourth terms and yields a corresponding growth timescale which is the sound crossing timescale of the perturbation (which for long wavelength perturbations is much larger than the thermal timescale). One can understand this result inasmuch as, in this case, it is the cooling at constant pressure which is destabilising, and hence one has to wait a timescale (equal to the sound-crossing timescale of the perturbation) for pressure balance with the surroundings to be re-established before the thermal instability can get under way. As k is increased, i.e. as the lengthscale of perturbations is reduced, the sound-crossing timescale decreases also, until it becomes comparable with the thermal timescale. At this point, the root of (10.93) corresponds to balance between the third and fourth terms and the growth timescale tends to the thermal timescale for all perturbations on this scale or smaller. Again, this can be readily understood, since once the sound-crossing timescale is less than the thermal timescale, then the re-establishment of pressure equilibrium is not the limiting process for the growth of the instability. On the other hand, when the cooling at constant pressure and at constant density are both destabilising ($B^* > 0$ and $A^* < 0$) then the growth timescale of the thermal instability is always roughly the thermal timescale, with the root of (10.93) at small and large k corresponding respectively to balance between the first and second, and third and fourth terms. This is because if the cooling at constant density is also unstable, then there is no requirement on perturbations to achieve pressure balance with the surroundings for the instability to grow.

The above analysis has made no assumption about the form of $\dot{Q}$. If we now return to the expression

$$\dot{Q} = A\rho T^\alpha - C, \tag{10.96}$$

which we can write as

$$\dot{Q} = \frac{A p \mu}{\mathcal{R}_*} T^{\alpha-1} - C, \tag{10.97}$$

then, for this restrictive form of the heating and cooling law, our stability criterion becomes

$$\left.\frac{\partial \dot{Q}}{\partial T}\right|_p = (\alpha - 1)\frac{A p \mu}{\mathcal{R}_*} T^{\alpha-2} \tag{10.98}$$

$\Rightarrow$

$$\boxed{\alpha \geq 1 \text{ for stability.}}$$

So optically thin thermal bremsstrahlung, $\alpha = 0.5$, is unstable.

Thermal instability forms a vital part of explaining the temperature structure of the interstellar medium in the Galaxy. Warm (temperature $T \sim 10^4$ K) and cold ($T \sim 10^2$ K) neutral atomic phases are known to co-exist. A hotter phase, ionised by old supernova remnants and hot stars, is also present.

Multi-phase equilibria are possible only if the gas can be thermally unstable. If there is only one temperature at which heating and cooling can be in equilibrium, then, for a given pressure, there will be only one phase. If more than one equilibrium is possible at a given pressure, then at least one of them will be thermally unstable. If the gas starts out in the unstable equilibrium phase, then it breaks up to form denser clouds which have lower temperatures than the initial medium, along with lower density regions of higher temperature in pressure equilibrium with the cool clouds. For a fuller description see the article by C. F. McKee in *The Physics of the Interstellar Medium and Intergalactic Medium*, ASP Conference Series, Vol. 80, 1995.

10.5 Method summary

The general approach to stability criteria through linearised analysis should now be second nature:

- Write down the equations.
- Decide on the equilibrium which is to be perturbed.
- Write the variables as equilibrium + Lagrangian perturbation, i.e. Δvariable.
- Calculate corresponding expressions for the Eulerian perturbations.
- Substitute expressions for the Eulerian perturbations (in terms of Δvariables) into the Eulerian fluid equations and retain terms only of order Δ.
- Write all Δs in the form coefficient times $e^{i\mathbf{k}\cdot\mathbf{x}+qt}$.
- Substitute into the equations and eliminate the coefficients.
- End up with a dispersion relation k in terms of q, where instability $\Leftrightarrow q$ real and > 0.

Chapter 11
Viscous flows

11.1 Linear shear and viscosity

The equation of continuity is valid for any fluid, since it expresses the conservation of mass. However, in a viscous fluid the old momentum equation has to be modified to take account of the transfer of momentum between fluid cells due to viscous processes. In Chapter 2 we derived the fluid equations, and expressed them in Cartesian form as:

$$\frac{\partial \rho}{\partial t} + \partial_j(\rho u_j) = 0 \tag{11.1}$$

and (2.21)

$$\frac{\partial(\rho u_i)}{\partial t} = -\partial_j \sigma_{ij} + \rho g_i, \tag{11.2}$$

where $g_i = -\partial_i \Psi$ in Cartesian coordinates.

We also had

$$\sigma_{ij} = \rho u_i u_j + p \delta_{ij}. \tag{11.3}$$

This equation was based on the premise that in the frame of a moving fluid element, its momentum does not change as a result of the differential motion of neighbouring fluid elements. Thus the term involving velocities in σ_{ij} above is just the result of momentum being advected *with* the fluid (i.e. it appears when we cast the equation in Eulerian form, as above), whereas the second term represents a force on the element due to thermal pressure differentials, *not* due to velocity gradients in the flow. However, the equation needs to be modified if there is any process which can transfer the momentum associated with velocity differences between the elements. We call such processes *viscous* processes.

In this case, the stress tensor must be modified so as to include an extra term, i.e. we write

$$\sigma_{ij} = \rho u_i u_j + p \delta_{ij} + \sigma'_{ij}, \tag{11.4}$$

where σ'_{ij} is the *viscous stress tensor*.

We shall shortly establish the form of the viscous stress tensor from general arguments. However, it is first instructive to consider a physical model for viscous transport processes at a microscopic level and derive the form of the viscous stress tensor in this case.

For this simple thought experiment, we just consider *shear* viscosity in a *linear shear flow*. In other words we set up a flow with parallel streamlines in which the (uniform) velocity gradient is perpendicular to the direction of the streamlines. This is illustrated in Figure 11.1 where the streamlines are in the i direction and the velocity gradient is in the j direction. If, at a microscopic level, the particles all travelled in the i direction, then there would be no communication between the different streamlines and the viscosity would be zero (i.e. the streamlines could slip freely past each other). However, if one considers that, due to the thermal motion, the particles have a random motion component in the j direction as well, then there is the possibility of communication between streamlines. Note that the ordinary fluid equations do not include such random motions because they just ascribe to any fluid element the *mean* velocity of all the particles in that element.

If we consider the flux of particles carrying an average momentum in the i direction corresponding to the velocity u_i across the surface with normal vector $\hat{e}_j$, and suppose the typical velocity of those particles in that direction is v_j relative to the bulk velocity, then the momentum flux from the fluid element in that direction is $\rho u_i v_j$. But we know that a typical particle (root-mean-square) velocity in a particular direction in a medium with temperature T is $\sqrt{\frac{kT}{m}}$, where k is Boltzmann's constant and m is the mass of each particle, so the appropriate value of v_j is $\alpha \sqrt{\frac{kT}{m}}$, where α will be some number of order unity, which would be obtained, in a detailed analysis, through appropriate averaging over

Fig. 11.1.

the directions of the random velocity vectors. (Note that in the case of momentum transport due to the thermal motions of the particles, as here, these velocity vectors should be isotropic in the local rest frame; we will later on consider the possibility of momentum transport through the bulk motion of turbulent fluid cells but note that in this case the random turbulent motions are not necessarily isotropic.)

In the element on the other side of the surface a similar situation applies, with an i-momentum flux across the surface now of $-\rho u_i^* \alpha \sqrt{\frac{kT}{m}}$, where the u_i^* etc. is the value of the streaming velocity in this element. Then, for a path length $\delta\ell$ we have

$$u_i^* = u_i + \partial_j u_i \, \delta\ell, \qquad (11.5)$$

so the net momentum flux in the j direction due to non-fluid processes (i.e. microscopic random motions) is

$$-\delta\ell \, \rho(\partial_j u_i)\alpha\sqrt{\frac{kT}{m}}, \qquad (11.6)$$

where we have considered the simple case that the flow is of constant density and temperature. In this case, the net momentum flux is

$$-\rho(\partial_j u_i)\alpha\sqrt{\frac{kT}{m}}\delta\ell. \qquad (11.7)$$

We now need to estimate $\delta\ell$, the scale length over which the momentum is transferred. We note that this momentum transfer will occur over a lengthscale comparable to the particle mean free path, since when particle–particle interactions occur, any momentum difference is redistributed between them. So an appropriate value for ℓ is the particle interaction cross-section divided by the particle number density. The particle number density $n = \rho/m$, so, treating the particles as hard spheres of radius a,

$$\ell \sim \frac{1}{\pi a^2 n} = \frac{m}{\pi a^2 \rho}. \qquad (11.8)$$

Therefore the net momentum flux is

$$-\frac{\alpha}{\pi a^2}\sqrt{mkT}\,\partial_j u_i. \qquad (11.9)$$

All we have to do now is to put this in the momentum equation for the fluid element, which becomes

$$\frac{\partial(\rho u_i)}{\partial t} = -\frac{\partial}{\partial x_j}(\rho u_i u_j + p\delta_{ij}) + \frac{\partial}{\partial x_j}\left(\eta\frac{\partial}{\partial x_j}u_i\right), \qquad (11.10)$$

where we have set

$$\eta = \frac{\alpha}{\pi a^2} \sqrt{mkT}. \tag{11.11}$$

We might guess that $\alpha \sim \frac{1}{2}$ (i.e. that half the particles going with a typical speed in the j direction are going in the opposite direction to the one towards the transfer surface we are considering), and then the shear viscosity coefficient $\eta \sim \frac{1}{2\pi a^2} \sqrt{mkT}$. This compares with a more rigorous derivation for the viscosity of dilute gases of hard spheres, where $\eta = \frac{5\sqrt{\pi}}{64\pi a^2} \sqrt{mkT}$. Given the approximations we have made, this is not bad agreement.

Note that the viscosity is independent of the density of the gas. This is something of a surprise, since we are used to denser materials being more viscous, but it has been confirmed by experiment. The denser gas has more atoms to transport the momentum, but this is compensated by the reduced mean free path over which the momentum transport can take place.

The other point is that the viscosity of the gas increases with temperature, since the rate at which the momentum can be transferred increases.

If the system is isothermal, then for an ideal gas the viscosity coefficient, η, is a constant. It is often taken outside the differential for this reason, even under circumstances when it is not strictly constant.

11.2 Navier–Stokes equation

We have seen that the viscous stress tensor, by definition, must involve terms that result from velocity differentials between fluid elements, i.e. terms proportional to $\frac{\partial u_i}{\partial x_j}$ for various values of i and j. We will start by writing the viscous stress tensor in the form

$$\sigma'_{ij} = -\eta \left(\frac{\partial u_i}{\partial x_j} + \frac{\partial u_j}{\partial x_i} - \frac{2}{3} \delta_{ij} \frac{\partial u_k}{\partial x_k} \right) - \zeta \delta_{ij} \frac{\partial u_k}{\partial x_k}, \tag{11.12}$$

where η and ζ are independent of the velocity. At first sight, this seems extraordinarily arbitrary! However, we shall see that, when written in this way, it has certain important and necessary properties. Firstly, it should be noted that the tensor is symmetric, which we can envisage as saying that the force on the ith face of a small cube in the jth direction is equal to the force on the jth face in the ith direction. Were this not true, then these two unbalanced forces would produce a torque on the cube. In the limit that the cube is very small (i.e. of mass tending to zero), this finite torque would impart infinite angular acceleration! This is avoided if we require that the viscous stress tensor is symmetric. (Although we have required that an infinitesimal fluid element should have zero torque acting on it, note that a larger cube of fluid *can* be torqued up or down by viscous stresses (see Section 11.3), since in this case the velocity field may vary over the region so that the forces on the side of the cube do not cancel. This situation is different from that of a flow involving only pressure forces since these act only normally on the surface and thus can *never* torque up or down a fluid element: see Helmholtz's equation (Chapter 9) in the case of inviscid flows.)

Secondly, this form for the viscous stress tensor ensures that all the diagonal terms $\propto \zeta$ are equal. What this means is that if the divergence of **u** is non-zero, and so the small fluid element is squashed (say), then the normal forces on the surfaces of the element, which resist such squashing, are the same on all surfaces. This is a defining property of an isotropic substance. There are many examples of solid substances which do not possess this particular attribute, because at a microscopic level the organisation of atoms/molecules has certain symmetries which mean that some directions experience larger stresses than others. Fluids, of course, do not possess any internal symmetries and are hence isotropic substances. (Again, we stress that unmagnetised fluids must be isotropic with respect to their microscopic structure, and it is therefore correct to impose this condition on the viscous stress tensor for *molecular* viscosity. In the case that viscosity arises from macroscopic 'turbulent' motion, the structure of turbulent eddies need not be isotropic and hence the viscous stress tensor need not necessarily be of this form.)

Once we have required that the tensor is symmetric, and that the diagonal terms $\propto \zeta$ are all equal, then the above expression is actually perfectly general, i.e. we can adjust the values of η and ζ to reproduce any desired dependence of the viscous stress through any surface on the various components of the velocity gradient. The reason why it is written in the form above, rather than defining a new coefficient (equal to $\zeta - 2/3\eta$) for the $\frac{\partial u_k}{\partial x_k}$ terms, is that, in the form above, the first term ($\propto \eta$) makes a zero contribution to the diagonal part of the stress tensor. Hence, as written, the term $\propto \eta$ is associated with momentum transfer in *shear flows*, whereas the second term ($\propto \zeta$) is associated with momentum transfer due to bulk compression of the flow. (To see why the latter is the case, note that $\frac{\partial u_k}{\partial x_k}$ is just $\nabla \cdot \mathbf{u}$, and therefore proportional to the rate of change of density locally.)

As a sanity check on the above derivation, we consider now the particular case of a flow in solid body rotation, i.e. where the flow velocity is given by $\mathbf{\Omega} \wedge \mathbf{r}$ for constant angular velocity $\mathbf{\Omega}$. Using the index representation of the cross product we have:

$$u_i = \epsilon_{ijk}\Omega_j x_k, \qquad (11.13)$$

where ϵ_{ijk} is the permutation symbol ($= 1$ if ijk is an even permutation of 123, -1 if an odd permutation, and zero if any of the indices are the same).

So

$$\frac{\partial u_i}{\partial x_j} = \epsilon_{ikl}\Omega_k \frac{\partial x_l}{\partial x_j} = \epsilon_{ikl}\Omega_k \delta_{lj} = \epsilon_{ikj}\Omega_k. \qquad (11.14)$$

Similarly we know

$$\frac{\partial u_j}{\partial x_i} = \epsilon_{jki}\Omega_k, \qquad (11.15)$$

and so

$$\frac{\partial u_i}{\partial x_j} + \frac{\partial u_j}{\partial x_i} = 0. \qquad (11.16)$$

We can also readily see that $\frac{\partial u_i}{\partial x_i}$ is zero for such a flow, so that the remaining terms in the viscous stress tensor are zero. We have therefore demonstrated that there is no viscous stress in the case of a flow in solid body rotation. This is exactly as we would expect, given that there is no shear (i.e. no slippage of fluid elements past each other) in a flow rotating as a solid body.

η and ζ are known respectively as the coefficients of shear and bulk viscosity for the fluid. We have set the signs in front of η and ζ in order that these coefficients are positive. The rationale for doing this in the case of the bulk term is obvious since with this definition the direction of momentum transfer is down the velocity gradient. We might also expect that in a shear flow the effect of viscosity should be to transfer momentum from fast to slow moving material, although the reason for this is perhaps not obvious until one considers how viscosity operates on a microscopic level (i.e. in terms of 'messenger' particles passing between the streamlines and carrying the specific momentum of their parent streamline; see Section 11.1 above). In fact we shall show below that – irrespective of the physical mechanism – shear viscosity must always transfer momentum in this direction (i.e. η must be positive as defined above) since otherwise it would violate the second law of thermodynamics by converting heat into kinetic energy.

Now we have a form for σ', the momentum equation is

$$
\frac{\partial(\rho u_i)}{\partial t} = -\partial_j \rho u_i u_j - \partial_j p \delta_{ij}
$$

$$
+ \partial_j \left[\eta \left(\frac{\partial u_i}{\partial x_j} + \frac{\partial u_j}{\partial x_i} - \frac{2}{3} \delta_{ij} \frac{\partial u_k}{\partial x_k} \right) + \zeta \delta_{ij} \frac{\partial u_k}{\partial x_k} \right] + \rho g_i, \qquad (11.17)
$$

or, using the continuity equation to remove the $\frac{\partial \rho}{\partial t}$ term,

$$
\rho \left(\frac{\partial u_i}{\partial t} + u_j \frac{\partial u_i}{\partial x_j} \right) = - \frac{\partial p}{\partial x_i}
$$

$$
+ \frac{\partial}{\partial x_j} \left[\eta \left(\frac{\partial u_i}{\partial x_j} + \frac{\partial u_j}{\partial x_i} - \frac{2}{3} \delta_{ij} \frac{\partial u_k}{\partial x_k} \right) \right]
$$

$$
+ \frac{\partial}{\partial x_i} \left(\zeta \frac{\partial u_k}{\partial x_k} \right) + \rho g_i. \qquad (11.18)
$$

This is the general form of the *Navier–Stokes* equation. η and ζ are usually functions of the temperature and density, so in general their spatial derivatives are not zero. However, in many cases it is an adequate approximation to treat η and ζ as constants, and so they are sometimes taken outside the spatial derivatives.

Bulk viscosity can usually be neglected in astrophysical fluids, the important exception to this being in shocks. We alluded a number of times in Chapter 7 to the importance of viscosity at the shock front, without any detailed explanation of what viscosity actually *is*. Now that we have considered this question more deeply, we see that in a shock (perpendicular to the x direction, say) it is the existence of a decelerating term in the x direction proportional to the (large) local velocity gradient in the x direction that causes the fluid to pass from supersonic to subsonic flow. If we knew the value of ζ, we could estimate the time required to decelerate the flow and, given the mean speed in the shock, also get an estimate of the shock thickness. Hence if we are interested in the internal structure of the shock, then we need to consider bulk viscosity (and likewise numerical codes need to include bulk viscosity if they are to manifest shocks). However, in many applications we are instead only interested in the properties of the flow on either side of the shock; in this case, as we saw in Chapter 7, we do not need to know anything about the viscosity and can just treat the shock as a discontinuity subject to jump conditions imposed by the (inviscid) fluid equations.

Since bulk viscosity is of rather limited relevance in astrophysics, we will henceforth retain only the terms due to shear viscosity, in which case the momentum equation becomes

$$\frac{\partial \mathbf{u}}{\partial t} + \mathbf{u} \cdot \nabla \mathbf{u} = -\frac{1}{\rho}\nabla p - \nabla \Psi + \frac{\eta}{\rho}\left[\nabla^2 \mathbf{u} + \frac{1}{3}\nabla(\nabla \cdot \mathbf{u})\right]. \qquad (11.19)$$

We note that since η (the coefficient of shear viscosity) appears in the combination $\frac{\eta}{\rho}$, it is convenient to define a new quantity $\nu = \frac{\eta}{\rho}$, which is known as the *kinematic viscosity*.

This version of the Navier–Stokes equation is like the Euler momentum equation, but now with an extra term

$$\nu\left[\nabla^2 \mathbf{u} + \frac{1}{3}\nabla(\nabla \cdot \mathbf{u})\right]. \qquad (11.20)$$

However, the mathematical nature of the problem is transformed by the addition of this term, since there is now a higher order spatial derivative in the Navier–Stokes equation. Consequently, we have to provide more boundary conditions, at least for flows in finite regions, than in the zero viscosity case. For example, in the case of ideal (viscosity-free) problems then it is sufficient, for example, to specify as boundary conditions that the flow normal to a bounding surface is zero; when $\nu \neq 0$ we also have to specify the tangential component there.

11.3 Evolution of vorticity in viscous flows

If we take the curl of the Navier–Stokes equation, in the same way as we can for the Euler equation, then (remembering that the vorticity $\mathbf{w} = \nabla \wedge \mathbf{u}$) we find

$$\frac{\partial \mathbf{w}}{\partial t} + \nabla \wedge (\mathbf{u} \cdot \nabla \mathbf{u}) = \nabla \wedge \left(-\frac{1}{\rho}\nabla p - \nabla \Psi + \frac{\eta}{\rho}\left[\nabla^2 \mathbf{u} + \frac{1}{3}\nabla(\nabla \cdot \mathbf{u})\right]\right).$$
$$(11.21)$$

Now since curl of grad is zero, we can write

$$\mathbf{u} \cdot \nabla \mathbf{u} = \frac{1}{2}\nabla u^2 - \mathbf{u} \wedge (\nabla \wedge \mathbf{u}) = \frac{1}{2}\nabla u^2 - \mathbf{u} \wedge \mathbf{w}, \qquad (11.22)$$

and if the fluid is barotropic, then

$$\nabla \wedge \frac{1}{\rho}\nabla p = -\frac{1}{\rho^2}\nabla \rho \wedge \nabla p = 0 \qquad (11.23)$$

(because for a barotropic fluid, the gradients of p and ρ are parallel).

Therefore, after a little manipulation, we have

$$\frac{\partial \mathbf{w}}{\partial t} = \nabla \wedge (\mathbf{u} \wedge \mathbf{w}) + \nu \nabla^2 \mathbf{w}. \qquad (11.24)$$

Consequently Kelvin's vorticity theorem (vorticity is constant and moves with the fluid) no longer holds for viscous fluids, so vorticity may decay, or be produced, in a fluid flow. As we noted above, the creation or decay of vorticity in a viscous fluid is a result of the fact that viscous stresses do not act normally on surfaces within the fluid.

11.4 Energy dissipation in incompressible viscous flows

The presence of viscosity results in the dissipation of kinetic energy, which is transformed into heat. To illustrate this we consider energy dissipation for the incompressible case (where we do not have to worry about the $p\,\mathrm{d}V$ term so can see directly how the viscous term dissipates the kinetic energy, and helpfully some other simplifications occur).

The total kinetic energy in an incompressible fluid is

$$E_{\text{kin}} = \frac{1}{2}\rho \int u^2 \, \mathrm{d}V, \tag{11.25}$$

and

$$\frac{\partial(\frac{1}{2}\rho u^2)}{\partial t} = \rho u_i \frac{\partial u_i}{\partial t} = \rho u_i \left(-u_k \frac{\partial u_i}{\partial x_k} - \frac{1}{\rho}\frac{\partial p}{\partial x_i} - \frac{1}{\rho}\frac{\partial \sigma'_{ik}}{\partial x_k} \right) \tag{11.26}$$

from the Navier–Stokes equation. Hence

$$\frac{\partial(\frac{1}{2}\rho u^2)}{\partial t} = -\rho u_k \partial_k \left[\frac{1}{2}u^2 + \frac{p}{\rho} \right] + \partial_i u_k \sigma'_{ik} - \sigma'_{ik}\partial_k u_i. \tag{11.27}$$

(the last two terms have been obtained by swapping the indices i and k – which is allowed, since both are repeated indices – and by noting that $\sigma'_{ik} = \sigma'_{ki}$ for the reasons explained above).

Then, since we have assumed that the fluid is incompressible, $\partial_i u_i = 0$ and so the first term in the RHS becomes a divergence and Equation (11.27) becomes

$$\frac{\partial(\frac{1}{2}\rho u^2)}{\partial t} = -\partial_i \left[\rho u_i \left(\frac{1}{2}u^2 + \frac{p}{\rho} \right) + u_k \sigma'_{ik} \right] + \sigma'_{ik}\partial_k u_i. \tag{11.28}$$

The term in square brackets is just the energy density flux in the fluid. The $\rho \mathbf{u}\left(\frac{1}{2}u^2 + \frac{p}{\rho}\right)$ is the flux due to the transfer of fluid mass, and the other term in the square brackets is the energy flux due to processes of internal friction. We know that the presence of viscosity always results in a momentum flux σ'_{ik}. A transfer of momentum also involves a transfer of energy, with an energy flux equal to the scalar product of the momentum flux and the velocity.

We can integrate (11.28) over some volume V, to obtain

$$\frac{\partial}{\partial t} \int \frac{1}{2} \rho u^2 \, dV = -\oint \left[\rho \mathbf{u} \left(\frac{1}{2} u^2 + \frac{p}{\rho} \right) + \mathbf{u} \cdot \underline{\sigma}' \right] \cdot d\mathbf{S} + \int \sigma'_{ik} \frac{\partial u_i}{\partial x_k} \, dV,$$

$$(11.29)$$

where by $\mathbf{u} \cdot \underline{\sigma}'$ we mean the vector whose components are $u_k \sigma'_{ik}$. The surface integral arises because of the divergence theorem, and is the rate of change of kinetic energy of the fluid in V owing to the energy flux through the surface bounding V. (Alternatively, we can regard it as the rate of work done by the pressure – both ram and thermal – and the viscous stress over the surface.) The remaining volume integral is the rate of decrease in kinetic energy due to viscous dissipation (see below).

If we extend the domain of integration to cover the whole extent of the fluid, then the surface integral vanishes (either because the velocity across a bounding surface is zero, or because the velocity at ∞ can be taken as zero), so we are left with

$$\frac{\partial E_{\text{kin}}}{\partial t} = \int \sigma'_{ik} \frac{\partial u_i}{\partial x_k} \, dV = \frac{1}{2} \int \sigma'_{ik} \left(\frac{\partial u_i}{\partial x_k} + \frac{\partial u_k}{\partial x_i} \right) dV, \qquad (11.30)$$

since σ'_{ik} is symmetric. But for an incompressible fluid $\sigma'_{ik} = \eta \left(\frac{\partial u_i}{\partial x_k} + \frac{\partial u_k}{\partial x_i} \right)$ since $\frac{\partial u_k}{\partial x_k} = \nabla \cdot \mathbf{u} = 0$, so

$$\frac{\partial E_{\text{kin}}}{\partial t} = \int \sigma'_{ik} \frac{\partial u_i}{\partial x_k} \, dV = -\frac{1}{2} \eta \int \left(\frac{\partial u_i}{\partial x_k} + \frac{\partial u_k}{\partial x_i} \right)^2 dV. \qquad (11.31)$$

This result confirms our choice of sign for η. If η is positive (as assumed), mechanical (i.e. kinetic) energy decreases as a result of viscous processes. Implicitly, this kinetic energy lost is converted into heat (and maybe then radiated). If η were negative, the reverse would occur (i.e. kinetic energy would be created at the expense of thermal energy, in violation of the second law of thermodynamics). We note that the direction of energy flow (i.e. from kinetic energy to thermal energy) is *always* the same, regardless of the form of the velocity field. This energy transfer associated with viscosity is therefore an *irreversible* one and is often described as 'viscous dissipation'.

11.5 Viscous flow through a circular pipe and the transition to turbulence

To illustrate the use of the Navier–Stokes equation we consider the steady flow of an incompressible viscous liquid through a horizontal pipe of constant diameter circular cross-section. We take the axis of the pipe as the z axis, and the fluid flow is then in the z direction. The velocity is then independent of z (from the continuity equation) but will be a function of the other two coordinates. Under these circumstances the Navier–Stokes equation

$$\frac{\partial \mathbf{u}}{\partial t} + \mathbf{u} \cdot \nabla \mathbf{u} = -\frac{1}{\rho}\nabla p - \nabla \Psi + \frac{\eta}{\rho}\left[\nabla^2\mathbf{u} + \frac{1}{3}\nabla(\nabla \cdot \mathbf{u})\right] \qquad (11.32)$$

becomes

$$\mathbf{u} \cdot \nabla \mathbf{u} = -\frac{1}{\rho}\nabla p + \nu\nabla^2\mathbf{u}. \qquad (11.33)$$

Now since $\mathbf{u} = (0, 0, u)$, where u is a function of x and y, we can see that the LHS is zero, and so

$$\nabla^2\mathbf{u} = \frac{1}{\eta}\nabla p. \qquad (11.34)$$

Since the x and y components of the velocity are zero everywhere, $\frac{\partial p}{\partial x} = \frac{\partial p}{\partial y} = 0$, i.e. the pressure is constant over the cross-section of the pipe. The z component of the equation gives

$$\frac{\partial^2 u}{\partial x^2} + \frac{\partial^2 u}{\partial y^2} = \frac{1}{\eta}\frac{\partial p}{\partial z}. \qquad (11.35)$$

The left hand side is a function of x and y only, and, as we have just shown, p depends only on z, so $\frac{\partial p}{\partial z}$ is a function of z only. Therefore $\frac{\partial p}{\partial z}$ must be a constant, and we can replace it by the $\frac{-\Delta p}{\ell}$, where Δp is the pressure difference over the pipe of length ℓ.

So far the shape of the pipe's cross-section has not entered into the discussion – the equations apply for viscous flow through any pipe with constant cross-section. We now choose to make the pipe circular, and, since there is now no dependence on the ϕ coordinate, we have

$$\frac{1}{R}\frac{d}{dR}\left(R\frac{du}{dR}\right) = -\frac{\Delta p}{\eta\ell}. \qquad (11.36)$$

We can integrate this to get

$$u = -\frac{\Delta p}{4\eta\ell}R^2 + a\ln R + b. \qquad (11.37)$$

Since the velocity at the centre of the pipe is finite, a must be zero, and the boundary condition that $u = 0$ on the inner surface of the pipe (at $R = R_0$) then gives us the result that

$$u = \frac{\Delta p}{4\eta\ell}\left(R_0^2 - R^2\right), \qquad (11.38)$$

i.e. the velocity distribution across the pipe is parabolic.

Most people who deal with pipes do not really have a great interest in the velocity profile, but they do want to know how much liquid they can pump through them. The mass of fluid passing through an annular element $2\pi R\,\mathrm{d}R$ per unit time is $2\pi R\rho u\,\mathrm{d}R$, so the total mass flow rate is

$$Q = 2\pi\rho \int_0^{R_0} Ru\,\mathrm{d}R. \qquad (11.39)$$

Using (11.38) then gives us

$$Q = \frac{\pi\Delta p}{8\nu\ell} R_0^4. \qquad (11.40)$$

Thus the mass of the fluid is proportional to the fourth power of the pipe radius. Note that this is not a question we can address without the viscosity term. As $\nu = \frac{\eta}{\rho} \to 0$ the flow rate becomes infinite. Or, in an ideal (zero viscosity) incompressible flow, there is no pressure gradient along a uniform pipe.

This analysis applies in the case that the flow is regular, which happens if the flow is not very fast. If Δp is increased, it turns out that the flow becomes irregular in space and time ('turbulent'), and this solution ceases to apply. We do not discuss turbulence in any detail here, but note that its treatment involves the breaking up of all variables into mean and fluctuating parts and the derivation of new equations in terms of these variables.

We will however consider the circumstances under which the turbulent regime is entered. We have just stated that laboratory experiments indicate that the transition to turbulence occurs above a certain critical speed, and we now enquire whether we can use this fact to determine when astrophysical flows become turbulent. For this simple discussion we will ignore the fact that, unlike the laboratory experiment of flow down a pipe, most astrophysical flows are highly compressible and may be threaded by dynamically important magnetic fields – the subject of compressible magnetohydrodynamic turbulence is well beyond the scope of this book. For the simple incompressible case, we may note that it is found experimentally that the onset of turbulence also depends on the kinematic viscosity of the fluid and this leads one to the hypothesis that the flow may be classified according to the value of a dimensionless variable involving kinematic viscosity ν and velocity V. The most obvious dimensionless combination also involves a lengthscale L associated with the flow (in the present case, related to the diameter of the pipe). In fact it is found that such a combination (i.e. $\mathcal{R} = LV/\nu$, known as the *Reynolds number* of the flow) turns out to be very useful in classifying viscous flow problems, since flows with the same Reynolds number indeed turn out to behave as scaled versions of each other. Thus, for example, the transition to turbulence is associated with a given Reynolds number (about 3000), irrespective of the individual values of the parameters L, V and ν.

Since turbulence is then associated with fast, large scale and low viscosity conditions, it should be no surprise that many astrophysical flows (for which the Reynolds number is very high) fall into the turbulent category. Although we will not be treating turbulence in any detail, we note here that the circular shear flows that we consider in the next chapter (i.e. accretion discs) have very high Reynolds numbers and are not expected to exhibit laminar flow. We will discuss how turbulence in such discs may provide a mechanism for mixing material across streamlines and thus provide an effective viscosity in accretion discs.

Chapter 12
Accretion discs in astrophysics

By far the most important application of the Navier–Stokes equations in astrophysics is to the case of circular shear flows, also known (for reasons that will become obvious once we have derived their properties) as *accretion discs*.

Such flows are encountered in many astronomical environments where gas is in nearly circular orbit around a massive central object. The scales of these flows vary immensely according to the nature of the central objects involved, which range from planets to stars to supermassive black holes. Figure 12.1 shows an example of what is believed to be an accretion disc around a black hole at the centre of a galaxy. Although, as we shall see, there are many qualitative aspects of these flows that are similar in all cases, the manner in which such rotating shear flows are created differs according to the type of astronomical system involved. For example, in the case of the primordial solar nebula, from which the planets in our Solar System were formed, the flow inherited its angular momentum from the slowly rotating cloud core which collapsed to form the Solar System. In the case of discs around supermassive black holes in the cores of galaxies, the origin of the gas orbiting the hole is not entirely understood, but virtually any explanation one could think of – be it gas shed from stars in the inner galaxy, or debris from stars tidally shredded by the black hole, or the swallowing of a small satellite galaxy – would involve gas that possessed significant angular momentum with respect to the central black hole. Another context in which discs are commonly encountered – this time around neutron stars or white dwarfs – is in the case of close binary star systems, where the origin of the material in orbit is gas donated by a companion star. Figure 12.2 illustrates the situation (which we have already considered in Chapter 5 in relation

Fig. 12.1. A giant disc of cold gas and dust fuels a possible black hole at the core of the galaxy NGC 4261 (right panel). Hot gas is ejected along the disc axis from the vicinity of the black hole creating the radio jets which are shown in the left panel. (NASA/STScI)

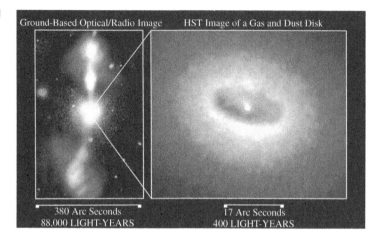

Ground-Based Optical/Radio Image   HST Image of a Gas and Dust Disk

380 Arc Seconds
88,000 LIGHT-YEARS

17 Arc Seconds
400 LIGHT-YEARS

Fig. 12.2. Schematic diagram of a close binary system containing a neutron star or white dwarf (on the right) and a giant star (on the left) which fills the critical equipotential and donates mass to its companion through the saddle point. The donated mass possesses angular momentum as a result of the binary's orbital motion and hence forms an accretion disc around the compact star.

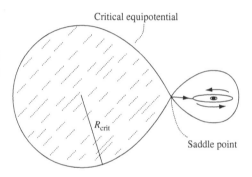

Critical equipotential

R_{crit}

Saddle point

to mass loss from the companion star): evidently material that crosses the critical equipotential separating the two stars possesses angular momentum with respect to the accretor star.

In each of these cases, therefore, we have set up a situation where we have gas with non-zero angular momentum that is bound to a central object. Whatever the origin of this material, and whatever its initial orbital trajectories, it will settle into a plane defined by the mean angular momentum vector of the gas supply. (Residual motion in other directions will be damped out on a free-fall timescale by shocks between colliding fluid elements, but once the gas has settled into a circular orbit, centrifugal force prevents its further radial collapse on this timescale.) Thus the net effect is that the gas settles into a ring-like, or disc-like, configuration.

It is an easy matter to demonstrate that such a system is a *shear flow*. (Note: a circular shear flow is one in which the *angular* velocity is a function of radius, just as a linear shear flow is one where the

linear velocity varies between streamlines.) For matter in circular orbit
at radius R around a mass M the centrifugal force and the gravitational
attraction balance, so

$$\Omega^2 R = \frac{GM}{R^2} \tag{12.1}$$

and hence

$$\Omega = \left(\frac{GM}{R^3}\right)^{\frac{1}{2}}. \tag{12.2}$$

This relation applies to the motion of planets around a star, and its
application in the Solar System is a restatement of Kepler's third law
of planetary motion. For this reason circular motion satisfying (12.2)
is referred to as *Keplerian motion*.

If we consider a gaseous disc where gravitational force from the
central star dominates, then the angular velocity will obey (12.2)
closely, and so evidently this is a shear flow (i.e. $\frac{d\Omega}{dR} \neq 0$). Due to
the action of viscosity, we expect the angular momentum to be transferred from the faster-moving inner regions to the slower-moving outer
regions of the disc. As the inner material loses angular momentum, it
moves inward on a spiral path. Therefore it is the viscosity which determines the rate at which the gravitational potential energy is converted
to other forms. Without viscosity, nothing happens – the elements of
the disc continue in circular orbits.

12.1 Derivation of viscous evolution equations for accretion discs

We choose cylindrical coordinates for obvious reasons. We expect u_ϕ
to be the dominant velocity component, with a small radial flow u_R as
a result of the effects of the viscosity. We assume $u_z = 0$, and $\frac{\partial}{\partial \phi} = 0$
(i.e. the disc is axisymmetric). Then the continuity equation is

$$\frac{\partial \rho}{\partial t} + \frac{1}{R}\frac{\partial}{\partial R}(R\rho u_R) = 0, \tag{12.3}$$

and the ϕ component of the Navier–Stokes equation (11.19) becomes,
in cylindrical polar coordinates,

$$\rho\left(\frac{\partial u_\phi}{\partial t} + u_R\frac{\partial u_\phi}{\partial R} + \frac{u_R u_\phi}{R}\right)$$
$$= \eta\left(u_\phi'' + \frac{1}{R}u_\phi' - \frac{u_\phi}{R^2}\right) + \frac{\partial \eta}{\partial R}\left(u_\phi' - \frac{u_\phi}{R}\right) \tag{12.4}$$

where primes denote differentiation with respect to R.

Note that here we have used the more general form of the Navier–Stokes equation, so have not assumed that η is a constant. The right hand side of Equation (12.4) is obtained by evaluating the right hand side of Equation (11.19) in cylindrical polar coordinates (see Appendix), with bulk viscosity set to zero.

The first thing we want to do is remove any z terms by integrating the equations through the depth of the disc. This will implicitly neglect any variation of u_R and u_ϕ with z, but any such dependence will be small. If we set the surface density $\Sigma = \int \rho \, dz$, then (12.3) becomes

$$\frac{\partial \Sigma}{\partial t} + \frac{1}{R} \frac{\partial}{\partial R} (R \Sigma u_R) = 0, \tag{12.5}$$

and (12.4) becomes

$$\Sigma \left(\frac{\partial u_\phi}{\partial t} + u_R \frac{\partial u_\phi}{\partial R} + \frac{u_R u_\phi}{R} \right) = \nu \Sigma \left(u''_\phi + \frac{1}{R} u'_\phi - \frac{u_\phi}{R^2} \right) + \frac{\partial \nu \Sigma}{\partial R} \left(u'_\phi \frac{u_\phi}{R} \right). \tag{12.6}$$

where primes denote differentiation with respect to R.

Note that on integrating over z the term involving $\frac{\partial^2 u_\phi}{\partial z^2}$ will vanish, on the assumption that $\eta \frac{\partial u_\phi}{\partial z}$ vanishes on the top and bottom surfaces of the disc. Using $\eta = \nu \rho$, if we now take (12.5)$\times R u_\phi$ plus (12.6)$\times R$, with $\Omega = u_\phi / R$, we have

$$\frac{\partial}{\partial t} (R \Sigma u_\phi) + \frac{1}{R} \frac{\partial}{\partial R} (\Sigma R^2 u_\phi u_R) = \frac{1}{R} \frac{\partial}{\partial R} \left(\nu \Sigma R^3 \frac{d\Omega}{dR} \right). \tag{12.7}$$

Here ν is strictly the density weighted average value of ν over z.

We can readily see the physical significance of this equation by noting that the first and second terms on the left hand side are respectively the (Eulerian) rate of change of angular momentum per unit area of an annulus at R and the net rate of angular momentum loss from this unit area due to advection of angular momentum with the radial flow. Any imbalance between these terms implies that the angular momentum content of this region is evolving as a result of a net (viscous) torque on the region, and this is therefore what the right hand side must represent.

As a sanity check, we multiply both sides of the equation by $2\pi R \, dR$ (so that we are now considering the rate of change of angular momentum in an annulus of width dR) so that the new right hand side ($2\pi \, dR \frac{\partial}{\partial R} (\nu \Sigma R^3 \frac{d\Omega}{dR})$) should now represent the *net* torque on the annulus. (Recall that the annulus should experience a spin-up torque as a result of its viscous interaction with more rapidly rotating material at smaller radius and a spin-down torque due to its viscous interaction with slower-moving material at larger radius. It only changes its angular momentum if there is an imbalance between these two.) But the *net* torque should be $dR\frac{dG}{dR}$, where G is the torque at R, and hence we deduce that (12.7) is telling us that

$$G(R) = 2\pi\nu\Sigma R^3 \frac{d\Omega}{dR}. \qquad (12.8)$$

We can check this by recalling that the torque at R is just the viscous force times R, where the viscous force is the product of the area and the viscous stress. The relevant area is the 'side' of the annulus where it 'rubs' against the adjoining annulus, i.e. $2\pi \times 2HR$ where H is a measure of the disc height, such that we can write $\Sigma = 2H\rho$. But if we look at the viscous stress tensor in cylindrical coordinates, we see that the relevant component of this tensor, $\sigma_{r\phi}$ (i.e. the force per unit area in the tangential direction acting on a surface with its normal in the radial direction), is $\rho\nu R\frac{d\Omega}{dR}$, i.e. $\eta R\frac{d\Omega}{dR}$. If we put these quantities together, we thus expect G to be $2\pi\nu\Sigma R^3 \frac{d\Omega}{dR}$, i.e. consistent with (12.8).

We have therefore argued in a loop, but a loop that has hopefully illuminated what is happening physically. If we apply the Navier–Stokes equations 'blindly', we get Equation (12.7). We then realise that the right hand side of this equation must relate to the net torque across an annulus, and can check what this must be by looking at the relevant part of the viscous stress tensor. Of course, we find consistency since the Navier–Stokes equations are derived from the expression for the viscous stress in Equation (11.12).

The behaviour of a viscous disc is then governed by Equation (12.7) and the continuity equation (12.5). Now is a good time to make a further approximation, i.e. that the material in the disc is in nearly Keplerian orbits. Then we can write

$$\Omega = \left(\frac{GM}{R^3}\right)^{\frac{1}{2}} \qquad (12.9)$$

and substitute in the above two equations and eliminate u_R. After a bit of manipulation the result is

$$\frac{\partial\Sigma}{\partial t} = \frac{3}{R}\frac{\partial}{\partial R}\left[R^{\frac{1}{2}}\frac{\partial}{\partial R}\left(\nu\Sigma R^{\frac{1}{2}}\right)\right], \qquad (12.10)$$

which describes accretion disc evolution.

12.2 Viscous evolution equation with constant viscosity

If ν is a constant, or a function of R alone, then one can solve this by separation of variables. Such solutions are interesting in that they give us a flavour of what is going on, but should not be treated as providing physically based predictions.

If we take ν as a constant, and $s = 2\sqrt{R}$, then (12.10) becomes

$$\frac{\partial}{\partial t}(R^{\frac{1}{4}}\Sigma) = \frac{12\nu}{s^2}\frac{\partial^2}{\partial s^2}(R^{\frac{1}{4}}\Sigma).$$
(12.11)

Hence we can write $R^{\frac{1}{4}}\Sigma = T(t)S(s)$, and this gives us

$$\frac{T'}{T} = \frac{12\nu}{s^2}\frac{S''}{S} = \text{constant} = -\lambda^2,$$
(12.12)

where the dashes denote derivatives with respect to the appropriate variables. Then the time dependence is exponential, and the space dependence is a Bessel function.

It is interesting to find the Green's function which is, by definition, the solution for $\Sigma(R, t)$ taking the initial mass distribution as a delta-function. So, if initially we have mass m at R_0,

$$\Sigma(R, 0) = \frac{m}{2\pi R_0}\delta(R - R_0),$$
(12.13)

then using dimensionless variables $x = R/R_0$, $\tau = 12\nu t R_0^{-2}$, the result is

$$\Sigma(x, \tau) = \frac{m}{\pi R_0^2 \tau x^{\frac{1}{4}}}\exp\left(-\frac{1+x^2}{\tau}\right)I_{\frac{1}{4}}(2x/\tau).$$
(12.14)

The $I_{\frac{1}{4}}(2x/\tau)$ is a modified Bessel function. Figure 12.3 illustrates the spreading of a viscous ring according to Equation (12.14).

The action of viscosity on the ring is to spread it out. Since time enters Equation (12.14) only via the combination $\tau = 12\nu t R_0^{-2}$, we see that the characteristic timescale for spreading of a ring of radius R_0 is $t_{\text{visc}} \sim R_0^2/\nu$. Most of the mass moves inwards, losing energy and angular momentum as it does so, but a small amount of matter moves out to take up the angular momentum lost by the material which spirals in. We can also see this from the asymptotic behaviour of u_R. In general

$$u_R = -3\nu\frac{\partial}{\partial R}\ln(R^{\frac{1}{4}}\Sigma)$$
(12.15)

and with the solution we have in the dimensionless variables this becomes

$$u_R = -\frac{3\nu}{R_0}\frac{\partial}{\partial x}\left[\frac{1}{4}\ln x - \frac{(1+x^2)}{\tau} + \ln I_{\frac{1}{4}}(2x/\tau)\right].$$
(12.16)

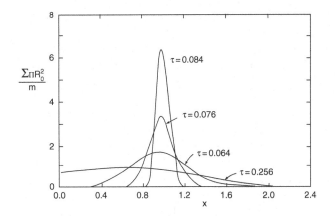

Fig. 12.3. The viscous evolution of a ring of matter of mass m. The surface density Σ is shown as a function of dimensionless radius $x = R/R_0$, where R_0 is the initial radius of the ring, and of dimensionless time $\tau = 12\nu t/R_0^2$ where ν is the viscosity. (Reprinted, with permission, from the *Annual Review of Astronomy & Astrophysics*, Volume 19 © 1981 by Annual Reviews www.annualreviews.org.)

The asymptotic behaviour of $I_{\frac{1}{4}}(z)$ is

$$I_{\frac{1}{4}}(z) \begin{cases} \propto z^{-\frac{1}{2}} e^z & \text{for } z \gg 1, \\ \propto z^{\frac{1}{4}} e^z & \text{for } z \ll 1. \end{cases} \tag{12.17}$$

Hence

$$u_R \sim \frac{3\nu}{R_0} \left[\frac{1}{4x} + \frac{2x}{\tau} - \frac{2}{\tau} \right] \quad \text{for } 2x \gg \tau, \tag{12.18}$$

which is > 0 for $x > 1$, and

$$u_R \sim -\frac{3\nu}{R_0} \left[\frac{1}{2x} - \frac{2x}{\tau} \right] \quad \text{for } 2x \ll \tau, \tag{12.19}$$

which is < 0 provided $\tau > 4x^2$. Hence the outer parts $(2x > \tau)$ move outwards and the inner parts inwards towards the accreting star. Also, the radius at which u_R changes sign moves steadily outwards, since if at some time equivalent to τ a given x is much greater than τ, if one waits long enough x will become $\ll \tau$. Consequently regions which initially move to larger radii subsequently lose angular momentum to material even further out and drift inwards. So at very long times almost all the mass m has accreted onto the star, and all the original angular momentum has been carried out by a very small fraction of the mass. This behaviour is apparent in Figure 12.3 where the centre of mass of the ring is evidently moving to smaller radii with time.

12.2.1 Comment on the viscosity in accretion discs

We derived above that the radial velocity due to viscous flow is of order ν/R and thus that the timescale for radial flow is $t_\nu \sim R^2/\nu$.

In fact we can obtain this as a general result (to order of magnitude) in viscous flows by considering the viscous torque (Equation (12.8)) which (omitting all numerical constants) we can write as $G \sim \nu \Sigma R^3 \frac{d\Omega}{dR}$. The net torque across an annulus of width ΔR is, in the same spirit, $G \times \Delta R / R$, where we have assumed that G varies over a lengthscale of order R, so that we can write $\frac{dG}{dR} \sim G/R$. But this net torque can be equated with the angular momentum content of the annulus $\sim \Sigma R \, dR R^2 \Omega$ divided by the timescale (t_ν) on which the material in the annulus flows in (or out) over a radial distance R. Again, setting $\frac{d\Omega}{dR} \sim \Omega/R$, we recover the above result that $t_\nu \sim R^2/\nu$ in general.

We may rewrite this timescale as $t_\nu \sim R/u_\phi \times R u_\phi/\nu = R/u_\phi \times \mathcal{R}$, where $\mathcal{R}$ is the Reynolds number, which we discussed in Chapter 11 in the context of flow down a pipe. Therefore we have derived the important result that in an accretion disc *the timescale for radial evolution due to viscosity exceeds the local orbital timescale by a factor of order the Reynolds number of the flow.*

Now let us make the provisional assumption that the viscosity mechanism in accretion discs is just that due to random particle motions which we introduced in Section 11.1. In this case, $\nu \sim c_s \ell$ from Equations (11.8), (11.11) and the definition $\nu = \eta/\rho$, so that the Reynolds number ($\mathcal{R}$) for the accretion disc flow is then

$$\mathcal{R} \sim R u_\phi \sigma n/c_s, \tag{12.20}$$

where n is the number density of particles (i.e. number m^{-3}), and σ the cross-section for their interaction (typically 10^{-20} m^2). As an example, consider material in the primordial solar nebula orbiting at the present position of the Earth (i.e. at 1au $= 1.5 \times 10^{11}$ m from the Sun). Material at that radius from the young Sun would have a temperature of several hundreds of degrees (similar to, but somewhat hotter than present conditions on Earth). For the density, we can 'grind up' all the mass in the planets, add in a correction for the hydrogen gas that has been lost from the system, and place it in an equilibrium disc with radius comparable with the distance of the Sun to Pluto. We also assume a thickness to radius ratio for the disc of about 10%, this being appropriate to hydrostatic equilibrium conditions in a disc of this temperature. This yields a density of $n \sim 10^{22}$ cm^{-3}.

Putting all this together, we obtain a very rough figure for the Reynolds number, which is enormous ($\sim 10^{14}$). From the relationship between viscous timescale, orbital time and Reynolds number above, we thus see that the timescale on which fluid would spiral into the Sun due to angular momentum transport by molecular viscosity is far greater than the age of the Universe. If we do a similar exercise in the other astrophysical contexts which we mentioned above (discs in Active Galactic Nuclei or in close binary stars) we get a similar answer: if molecular viscosity were the only viscous process operative, then discs would simply not accrete on any timescale that would make them observable. In other words, material would orbit the central object indefinitely at a fixed radius and there would be no mechanism for releasing its gravitational energy as the accretion luminosity that we observe in these systems.

In fact, the very high Reynolds number provides a clue as to the nature of the viscosity mechanism. For most laboratory fluids the critical Reynolds number at which turbulence sets in is somewhere in the range 10–10^3, so it is highly likely that the accretion flow is turbulent. Under these circumstances the momentum transfer process we have suggested will be overwhelmed by a larger scale process due to the turbulent mixing, so there is a turbulent viscosity $\nu_{\mathrm{turb}} \sim \ell_{\mathrm{turb}} v_{\mathrm{turb}}$, where v_{turb} and ℓ_{turb} are respectively the velocity and lengthscales associated with the turbulence. Then there is little more we can say than that ℓ_{turb} is likely to be less than the disc thickness, and shocks are likely to dissipate supersonic turbulence so $v_{\mathrm{turb}} < c_{\mathrm{s}}$. However, since ℓ_{turb} is much larger than the molecular mean free path, this implies that ν_{turb} far exceeds the viscosity provided by molecular collisions. Thus the viscous timescale for the spread of an accretion disc t_{visc} can be much smaller than the estimate based on molecular viscosity.

We obtain the best clues as to the actual magnitude of the viscosity required through studying the variations in luminosity from accretion discs in close binary stars. These studies have shown that the viscosity must be one or two orders of magnitude less than the maximum value for turbulent viscosity derived above. In the last decade, there has been considerable progress in studying magnetohydrodynamical turbulence in accretion discs and it would seem, from simulations, that a viscosity of the required magnitude can indeed be sustained in this way.

There is a further subtlety about the *direction* of angular momentum transport in circular shear flows that has caused considerable confusion in the literature. In Section 11.1 we showed that a simple, kinetic theory type argument can be used to illustrate the way that linear momentum is transported by random motions in a linear shear flow, and we would like to be able to present similar arguments for the circular case. When particles travel from one streamline to another due to their random motion, then they must conserve their angular momentum between collisions (since they are just moving ballistically in a central force field). This suggests that – just as random motions in a linear shear flow transport linear momentum from streamlines with high to low momentum – then in the circular case, angular momentum should be transported from streamlines of high to low specific angular momentum.

This heuristic answer contradicts that yielded by the Navier–Stokes equation, which instead states that angular momentum is transported down the gradient of angular *velocity* (not angular momentum). In a Keplerian disc, this distinction would even reverse the direction of evolution, since whereas angular velocity declines with increasing radius, the specific angular momentum increases outwards. A little thought tells us that the Navier–Stokes answer is in fact the correct one. For example, it gives us the intuitively correct result that there is no viscous evolution for systems in solid-body rotation (because no relative motion of adjoining fluid elements). We also know that the Navier–Stokes equations drive accretion flows along the energetically favourable route of spreading a ring both inwards and outwards, whereas our heuristic result would, by implication, cause a ring to become thinner and thinner, raising its energy in the process.

Clearly then, there is something flawed in the logic of the heuristic argument. It is of no help at all to maintain that in any case molecular viscosity is irrelevant to real accretion discs, since one should be able to construct a qualitatively correct kinetic theory argument for turbulent angular momentum transport, replacing the motion of individual atoms by those of turbulent eddies.

The resolution of this apparent paradox is seen most simply by considering random particle motions in the frame co-rotating with the local flow. In this frame, particles do not travel in straight lines between collisions but are instead curved by Coriolis force. Considering just particles that move outwards from a streamline as a result of random motion, it turns out that those with a random velocity component along the mean flow ('prograde particles') are bent into more radial trajectories, whereas those whose random velocity vector is retrograde are bent into more trailing trajectories. Consequently, the *flux* of prograde particles arriving at a surface at larger radius exceeds that of retrograde particles. By definition, the angular momentum of the prograde particles exceeds that of the retrograde particles.

This effect is sufficient to ensure the correct direction of angular momentum transport. Even though the *mean* specific angular momentum of the flow at the inner streamline is less than that at the surface at larger radius, the *time average* of the angular momentum brought by the particles is such as to exert a spin-up torque on the surface, simply because the flux of particles received at the surface is biased towards the prograde particles. Consequently, angular momentum is indeed transported outwards (i.e. from high to low angular velocity) as indicated by the Navier–Stokes equation.

12.3 Steady thin discs

In Section 12.2, we illustrated the way that accretion discs work by determining how a thin ring of fluid would spread (i.e. the Green's function solution). Another solution of astrophysical interest is the steady state solution, which presupposes that there is some source of fluid that maintains a steady accretion rate through the disc. For a *thin* disc, then, setting the time-dependent part equal to zero gives us

$$R\Sigma u_R = C_1 \tag{12.21}$$

from (12.5), and

$$\Sigma R^3 \Omega u_R - \nu \Sigma R^3 \frac{d\Omega}{dR} = C_2 \tag{12.22}$$

from (12.8), where C_1 and C_2 are constants. For a steady disc, the mass inflow rate $\dot{m}$ is $-2\pi R\Sigma u_R$, and it follows that this is a constant and

$$C_1 = -\frac{\dot{m}}{2\pi}. \tag{12.23}$$

To calculate C_2 we need to consider an inner boundary condition. Let us suppose that the accreting matter flows onto the surface of the star which has radius R_*. In a realistic situation, the star will be rotating with angular velocity less than that of a Keplerian orbit at its surface, i.e. less than the break-up speed, so

$$\Omega_* < \Omega_K(R_*). \tag{12.24}$$

In this case the angular velocity of the accreting material increases inwards until it starts to decrease towards Ω_* in a boundary layer near the surface of the star. Hence there exists a radius $R = R_* + b$ at which $\frac{d\Omega}{dR} = 0$. Then, if $b \ll R_*$, it can be shown that Ω is very close to the Keplerian value at the point where $\frac{d\Omega}{dR} = 0$. This then implies, from Equations (12.21)–(12.23), and Kepler's equation (12.2) to substitute for Ω, that

$$C_2 = -\frac{\dot{m}}{2\pi} R_*^2 \Omega = -\frac{\dot{m}}{2\pi} (GMR_*)^{\frac{1}{2}}. \tag{12.25}$$

Then putting this constant in Equation (12.22) gives us

$$\nu\Sigma = \frac{\dot{m}}{3\pi} \left[1 - \left(\frac{R_*}{R}\right)^{\frac{1}{2}}\right]. \tag{12.26}$$

This shows that the mass inflow rate $\dot{m}$ and the viscosity ν depend linearly on each other for a fixed surface density profile in the disc.

We can now use the energy loss equation (11.31) in order to calculate the rate of viscous dissipation of energy in such a disc. Note that (11.31) is specifically derived for the case of an *incompressible* flow and therefore omits two additional energy changes associated with (a) dissipation by bulk viscosity and (b) $p\,dV$ work. The former involves a rate of change of energy per unit volume $\sim \eta(\nabla \cdot \mathbf{u})^2$ and the latter $\sim p(\nabla \cdot \mathbf{u})$. In an axisymmetric two-dimensional flow, the only contribution to $\nabla \cdot \mathbf{u}$ is $\frac{1}{R}\frac{\partial}{\partial R}(Ru_R)$, which is $\sim \frac{u_R}{R} \sim \Omega^{-1}\mathcal{R}^{-1}$, where $\mathcal{R}$ is the Reynolds number. It turns out that the ratios of the contributions from bulk viscosity and $p\,dV$ work to that for shear viscosity (as contained in (11.31)) are respectively $\sim(H/R)^2$ (for a disc of thickness H) and $\mathcal{R}^{-2}$. Thus for thin, high Reynolds number flow we can neglect these contributions and consider only the right hand side of (11.31), i.e. a viscous dissipation of $\eta R^2 \left(\frac{d\Omega}{dR}\right)^2$. Integrating this over z gives the energy dissipated per unit area of the disc, which is

$$F_{\text{diss}} = \int \eta R^2 \left(\frac{d\Omega}{dR}\right)^2 dz = \nu\Sigma R^2 \left(\frac{d\Omega}{dR}\right)^2. \tag{12.27}$$

Now using Kepler's law for Ω, and $\nu\Sigma$ from (12.26), gives

$$F_{\text{diss}} = \frac{3GM\dot{m}}{4\pi R^3}\left[1 - \left(\frac{R_*}{R}\right)^{\frac{1}{2}}\right]. \tag{12.28}$$

The total energy emitted by an accretion disc is obtained by integrating this over the entire disc area, so

$$L = \int_{R_*}^{\infty} F_{\text{diss}} 2\pi R\, dR = \frac{GM\dot{m}}{2R_*}. \tag{12.29}$$

Since $-GM/R_*$ is the gravitational potential at the surface of the star, $GM\dot{m}/R_*$ is the rate of gravitational energy loss due to the inflow $\dot{m}$. What this tells us is that half of the energy is emitted from the accretion disc. The other half remains as kinetic energy just before the gas reaches the object, where it may be dissipated in a thin boundary layer.

The fact that half the energy available is emitted from the accretion disc is particularly significant for black holes, since for them the gravitational energy of the infalling mass is of the same order as the rest mass energy of the material. One can deduce this by a pseudo-Newtonian approach in which we assume the material spirals into the hole via a normal Newtonian disc that is terminated at a few times GM/c^2 from the black hole. Of course, a proper treatment of the inner regions of an accretion disc around a black hole requires a General Relativistic treatment. This simple estimate shows that black hole accretion is a highly efficient way of converting rest mass energy into radiation, and current estimates are that the process is $\sim 10\%$ efficient. Nuclear burning in the cores of stars, by contrast, liberates at most $\sim 0.7\%$ of the rest mass energy. It is therefore unsurprising that accretion onto black holes in quasars and other Active Galactic Nuclei is believed to provide a significant fraction of the luminosity in the Universe, being at the very least comparable with that provided by nuclear burning in stars.

A final point needs to be made about the energy balance in accretion discs. We have just demonstrated that the total luminosity of the disc is just the mass flow rate times the change in specific energy of material flowing in on Keplerian orbits from infinity to the stellar surface. One *cannot*, however, make the same argument about the *local* energy balance in the disc. Considering fluid spiralling in, at rate $\dot{m}$, from radius $R + dR$ to R, then the rate at which the total energy of the flow decreases in this annulus is $-\dot{m}\left[\frac{GM}{2(R+dR)} - \frac{GM}{2R}\right] \sim GM\dot{m}\,dR/2R^2$. (Recall that the factor 2 comes from the fact that we are considering the *total* energy, which we can simply relate to the potential energy of a Keplerian flow using the virial theorem.) Thus the rate of change of total energy per unit area is the above divided by the area $2\pi R\,dR$ of the annulus, i.e. $GM\dot{m}/4\pi R^3$. However, the rate of energy dissipated per unit area at radius R is given by (12.28). In other words, at large radii the energy dissipated by viscosity is *three times* that liberated by the release of gravitational energy of material spiralling inwards locally. At small radii, by contrast, the energy dissipated by viscosity is much less than that liberated locally by inflowing material, and indeed goes to zero as $R \to R_*$. Integration of the above expression for energy loss per unit area over the entire disc however yields the same answer as integration of (12.28) over area, so clearly energy is being conserved globally.

The resolution of this apparent paradox is to return to Equation (11.29). The rate of change of energy flowing through an annulus changes due to *both* internal viscous dissipation *and* work done by viscous stresses at the boundaries of the annulus. In order for a steady Keplerian flow to be maintained, it is necessary that the *sum* of these two terms is equal to $GM\dot{M}\,dR/2R^2$, but the partition of this energy between the two terms varies as a function of radius. At small radii (i.e. as $R \to R_*$), almost all the energy lost by inflowing material is the result of work done by viscous torques against the flow (i.e. by spin-down torques on the outer edge of each annulus). Very little is dissipated within the annulus and is thereby available as a source of luminosity. At large radii, however, the rate of energy dissipation within the annulus far exceeds what is available from the steady inflow; this means that the difference must be put back into the flow due to work done by spin-up torques (on the inner edge of each annulus). In essence, therefore, the viscous stresses mediate the location in the disc where the energy is dissipated as heat (and therefore radiated): their net role (in a steady flow) is to redistribute energy to large radii so as to achieve the steady state dissipation profile given by (12.28).

12.4 Radiation from steady thin discs

If each annulus of the disc is optically thick (so all radiation is thermalized), it radiates as a black body with a temperature $T_{\mathrm{eff}}(R)$ such that the emitted flux balances the viscous dissipation F_{diss}, then from (12.28) we require

$$
2\sigma_{\mathrm{SB}} T_{\mathrm{eff}}(R)^4 = \frac{3GM\dot{m}}{4\pi R^3}\left[1 - \left(\frac{R_*}{R}\right)^{\frac{1}{2}}\right],
\tag{12.30}
$$

where σ_{SB} is the Stefan–Boltzmann constant. The extra factor of 2 on the left hand side of (12.30) arises because radiation emerges from both sides of the disc. For $R \gg R_*$, we then recover the characteristic power law temperature profile of a steady state accretion disc: $T_{\mathrm{eff}} \propto R^{-\frac{3}{4}}$.

The emitted spectrum of such a disc is obtained by summing black body spectra over annuli at the appropriate temperature, weighting the contribution from each according to its area. Thus the radiated flux at frequency ν (not to be confused with the kinematic viscosity, which is also conventionally labelled ν) is given by

$$
F_\nu = \int_{R_*}^{R_{\mathrm{out}}} B_\nu\left(T_{\mathrm{eff}}(R)\right) 2\pi R\, \mathrm{d}R,
\tag{12.31}
$$

where B_ν is the Planck function and R_{out} is the outer radius of the disc.

$$
B_\nu(T) = \frac{2h}{c^2}\frac{\nu^3}{e^{\frac{h\nu}{kT}} - 1},
\tag{12.32}
$$

where c is the speed of light, h is Planck's constant and k is Boltzmann's constant. The form of the resulting spectrum is shown in Figure 12.4. At low frequencies $\nu < kT_0/h$, where T_0 is the temperature at the outer radius, the spectral shape is the sum of terms on the Rayleigh–Jeans tail of the Planck distribution, so is $\propto \nu^2$. At high frequencies the spectrum is controlled by the Wein cutoff of the hottest regions, i.e. those near R_*. At intermediate frequencies the spectrum approaches a power law $F_\nu \propto M\dot{m}\nu^{\frac{1}{3}}$ though the range of frequencies where this approximation applies is small unless the temperature at the outer radius is very much less than that at R_*.

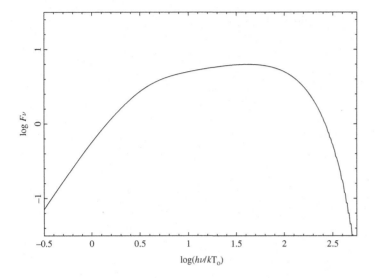

Fig. 12.4. The spectral energy distribution from an optically thick accretion disc with outer radius of $500R_*$. The $\nu^{\frac{1}{3}}$ power law slope for part of the spectrum is clearly seen near the centre of the frequency range shown. The flux scale is arbitrary.

Note that for a steady disc the form of F_ν from Equations (12.30) and (12.31) is independent of the form of the viscosity coefficient η (or the kinematic viscosity ν), and depends only on $\dot{m}$ (and of course M and the inner and outer disc radii). In some ways this is highly convenient, as the spectral signature of an optically thick disc is unambiguous, at least to this level of approximation. In reality radiative transfer effects in the disc atmospheres complicate the picture, particularly where there is strong continuum absorption (see Figure 12.5), but one can determine $M\dot{m}$ from such a spectrum. On the other hand, this fact is inconvenient

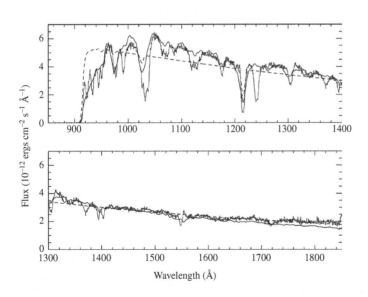

Fig. 12.5. The spectral energy distribution of the dwarf nova Z Camelopardalis in a quasi-steady state. The dashed line shows a black body model fit to the spectrum with hydrogen absorption overlaid to obtain the break at 912 Å. The solid curve is a model atmosphere fit. Note that the flux scale is F_ν, but it is plotted against wavelength $\lambda = c/\nu$. (From Knigge *et al.*, *Astrophysical Journal* **476**, 291, 1997)

if one wants to learn something about the viscosity in accretion discs, since all one learns is that the product of the kinematic viscosity and the surface density has somehow adjusted so as to deliver a particular $\dot{m}$.

As emphasised in Section 12.2, the only way to constrain the kinematic viscosity ν observationally is to study non-steady discs, since the timescale for radial redistribution of matter and consequent changes in luminosity is the viscous timescale $t_\nu \sim R^2/\nu$. Studies of outbursts from discs around young stars reveal that Reynolds numbers for these are in the range $\sim 10^4$–10^5. If the source of the viscosity is molecular then Reynolds numbers $\sim 10^{14}$ are expected. The viscosity in these discs must therefore be from some other processes which are incompletely understood, but are ten orders of magnitude more efficient than molecular viscosity.

Chapter 13
Plasmas

In this final Chapter, we introduce a new level of complexity in the physics of astrophysical fluids, i.e. we recognise that where fluids are composed of charged particles, there is the possibility that their behaviour is modified by electromagnetic fields. Such fluids are called plasmas. Here we will be able to do no more than introduce the fundamental principles of *magnetohydrodynamics* and select a few simple examples to illustrate some important applications in the interstellar medium.

Magnetic fields are important in many astrophysical situations. They may control the dynamics (e.g. in solar loops and flares, see Fig. 13.1) or the source of the radiation from small scales (e.g. pulsars) to very large scales (e.g. radio galaxies). A weak and largely disordered magnetic field with a strength of about 5×10^{-10} tesla permeates the interstellar medium of the galaxy. In general, observations show that the magnetic and kinetic energy of the interstellar medium are of similar magnitude, suggesting that magnetic and thermal processes may be closely coupled.

The relevance of magnetic fields for the motion of fluids consisting of charged particles arises because a charge moving with velocity $\mathbf{u}$ in a magnetic field $\mathbf{B}$ experiences a force proportional to $\mathbf{u} \wedge \mathbf{B}$. Thus the velocity component of the particle which is parallel to the magnetic field is unaffected, but the perpendicular velocity component gives rise to a force which is orthogonal to both the field direction and the perpendicular velocity direction. As a consequence, charged particles can move freely along magnetic field lines but their motion perpendicular to them is constrained. If the magnetic force dominates, then charged particles move in spirals around magnetic field lines.

Fig. 13.1. A coronal loop on the sun, extending to over 150,000 km above the photosphere. These are found around sunspots and active regions, and are associated with the closed magnetic field lines that connect magnetic regions on the solar surface. (NASA)

In general there will be an interplay between the magnetic field and the fluid of charged particles, with the magnetic field modifying the fluid motion and the fluid motions giving rise to changes in the magnetic field. Apart from having another variable to deal with, there is the additional complication that the magnetic forces on the particles in the fluid are not isotropic.

13.1 Magnetohydrodynamic equations

In gases in which the constituent particles are charged, and which satisfy the normal requirement that the overall charge is neutral, there are at least two species to keep track of (in the simplest case protons and electrons with charges $\pm e$). Obviously, these oppositely charged particles behave differently in electromagnetic fields.

At this stage we have a fundamental choice about how we treat the plasma. We could consider it as an ensemble of charged particles and derive statistical properties, such as distribution functions for particle velocities, for each species. This is somewhat analogous to the treatment of a galaxy full of stars in terms of distribution functions for the stellar orbits, except that it is more complicated in this case because of the additional physical interactions associated with particle charge. Alternatively, as in the case of stellar orbits, we can adopt a fluid approach, so that we instead derive mean properties (such as of particle speed) averaged over fluid elements. As in our discussion in Chapter 1, the latter approach requires that the problem allows the definition of fluid elements – relatively homogeneous regions over which one can average particle properties. What is now different is that in each fluid element, we can define different mean properties (e.g. density or velocity) for the particle species with different charges.

Unsurprisingly, in this book we are going to adopt the fluid approach, known as magnetohydrodynamics. We will consider a plasma consisting of two charged species: protons (mass m^+ per particle, charge e^+, and fluid velocity of the protons $\mathbf{u}^+$) and electrons (same quantities, with superscript minus). The extension to more species is obvious, and adds only to the tedium and nothing to the essentials.

We start with the usual conservation of matter equation for each species, which is the Eulerian equation with ρ replaced by the number density n, since the mass of each particle cancels from the equation. So

$$\frac{\partial n^+}{\partial t} + \nabla \cdot (n^+ \mathbf{u}^+) = 0$$

$$\frac{\partial n^-}{\partial t} + \nabla \cdot (n^- \mathbf{u}^-) = 0,$$

(13.1)

If we define a centre of mass velocity $\mathbf{u}$ by

$$\mathbf{u} = \frac{m^+ n^+ \mathbf{u}^+ + m^- n^- \mathbf{u}^-}{m^+ n^+ + m^- n^-},$$

(13.2)

then multiplying each of these by m^+ or m^-, as appropriate, gives us the usual mass conservation equation

$$\frac{\partial \rho}{\partial t} + \nabla \cdot (\rho \mathbf{u}) = 0,$$

(13.3)

where $\rho = m^+ n^+ + m^- n^-$.

We can similarly define the charge density $q = n^+ e^+ + n^- e^-$, and the current density $\mathbf{j} = e^+ n^+ \mathbf{u}^+ + e^- n^- \mathbf{u}^-$, and then equations (13.1) imply

$$\frac{\partial q}{\partial t} + \nabla \cdot \mathbf{j} = 0.$$

(13.4)

Integration of this equation over any volume and application of the divergence theorem shows us that this equation merely states that the rate of charge accumulation in any volume just depends on the currents in and out of the volume.

We then have to consider the momentum equation for the fluid, now with the Lorentz force acting on each particle due to the electric and magnetic fields $\mathbf{E}$, $\mathbf{B}$,

$$\mathbf{F} = e(\mathbf{E} + \mathbf{v} \wedge \mathbf{B}).$$

(13.5)

So, for the two species we have

$$m^+ n^+ \left(\frac{\partial \mathbf{u}^+}{\partial t} + \mathbf{u} \cdot \nabla \mathbf{u}^+ \right) = e^+ n^+ (\mathbf{E} + \mathbf{u}^+ \wedge \mathbf{B}) - f^+ \nabla p$$

$$m^- n^- \left(\frac{\partial \mathbf{u}^-}{\partial t} + \mathbf{u} \cdot \nabla \mathbf{u}^- \right) = e^- n^- (\mathbf{E} + \mathbf{u}^- \wedge \mathbf{B}) - f^- \nabla p,$$

$$(13.6)$$

where we have used unknown factors f^+ and f^- to allow for the fractions of the pressure gradient which go into accelerating each species. Note that the right hand side of each equation represents the forces acting on each particle species *within a given fluid element*; therefore the convective derivative is calculated using the mean velocity $\mathbf{u}$ of the element. We can now perform the sum over the two species, to obtain

$$\rho \left(\frac{\partial \mathbf{u}}{\partial t} + \mathbf{u} \cdot \nabla \mathbf{u} \right) = q\mathbf{E} + \mathbf{j} \wedge \mathbf{B} - \nabla p, \qquad (13.7)$$

since the sum of the fractions of the ∇p components in the individual species equations has to be unity.

To close this set of equations we need another relation for $\mathbf{j}$, and this comes from transforming Ohm's law for a conductor into a form appropriate to a frame in which $\mathbf{u}$ is non-zero. We have:

$$\mathbf{j} = \sigma(\mathbf{E} + \mathbf{u} \wedge \mathbf{B}) \qquad (13.8)$$

where σ is the electrical conductivity. Essentially, all this states is that due to the fluid velocity the magnetic field moves charges in addition to the electromotive force.

The significant change from the previous fluid equations is the term which gives the response of the particles to the presence of electric and magnetic fields, this being simply the sum of the Lorentz forces for all the constituent particles. In laboratory conditions, one can sometimes specify $\mathbf{E}$ and $\mathbf{B}$ in which the charged particles move as test particles (i.e. without much modifying the strong fields in which they move). In astrophysical plasmas, however, the usual situation is that the electric and magnetic fields are themselves generated by the motions and distributions of the charged particles and we therefore need additional equations to relate $\mathbf{E}$ and $\mathbf{B}$ to the charge and current distributions in the plasma. (This is equivalent to going from the study of the orbits of particles in a fixed gravitational potential to that of self-gravitating systems). In the electromagnetic case, the relationship between $\mathbf{E}$ and $\mathbf{B}$ and q and $\mathbf{j}$ is supplied by Maxwell's equations:

$$\nabla \cdot \mathbf{B} = 0 \qquad (13.9)$$

$$\nabla \cdot \mathbf{E} = \frac{q}{\varepsilon_0} \qquad (13.10)$$

$$\nabla \wedge \mathbf{B} = \mu_0 \mathbf{j} + \frac{1}{c^2} \frac{\partial \mathbf{E}}{\partial t} \qquad (13.11)$$

$$\nabla \wedge \mathbf{E} = -\frac{\partial \mathbf{B}}{\partial t} \qquad (13.12)$$

Any reader who is unfamiliar with these equations is directed to any electromagnetism text (e.g. Bleaney & Bleaney 'Electricity and Magnetism', volume 1, Oxford University Press, 1989). Briefly, the first is simply a statement that there are no sources or sinks of magnetic flux ('no magnetic monopoles'). By expressing the electric field as the gradient of a scalar potential, the second can be readily cast into a form analogous to Poisson's equation relating mass density to gravitational potential (see Chapter 3). The third equation relates the magnetic field to the current distribution: integration of this equation over area and application of Stokes theorem shows us its relation to simple steady state problems such as the evaluation of the field around an infinite straight current bearing wire (Ampere's law). The second term, involving the so-called displacement current, is crucial to the propagation of electromagnetic waves since, the third and fourth of Maxwell's equations imply that time dependent **B** fields imply the generation of **E** fields *and vice versa*. We shall shortly see, however, that it can usually be ignored in the case of non-relativistic plasmas. The fourth Maxwell equation is that of electromagnetic induction. Again, integration of this equation over area and application of Stokes' theorem relates the rate of change of magnetic flux crossing this area to the induced potential change across the surface, a result with obvious application to electrical dynamos.

13.2 Simplifying the magnetohydrodynamic equations

If we look at Maxwell's equations then the one we might prefer to simplify is (13.11), since it has two terms on the RHS and we might hope one of these would usually dominate. Since we are dealing with flows we might expect (or hope) that the $\frac{\partial \mathbf{E}}{\partial t}$ term in

$$\nabla \wedge \mathbf{B} = \mu_0 \mathbf{j} + \frac{1}{c^2} \frac{\partial \mathbf{E}}{\partial t} \qquad (13.13)$$

could be ignored.

The other equation linking $\mathbf{B}$ and $\mathbf{E}$ directly is equation (13.12):

$$\nabla \wedge \mathbf{E} = -\frac{\partial \mathbf{B}}{\partial t} \tag{13.14}$$

From this we would expect $E/B \sim \ell/\tau$, where ℓ and τ are the length- and time-scales for the fields. Since it is the interaction of the fluid flow and the fields which is important in MHD, we expect $\ell/\tau \sim u$, the velocity scale for the flow. Hence

$$\left| \frac{1}{c^2} \frac{\partial \mathbf{E}}{\partial t} \right| / |\nabla \wedge \mathbf{B}| \sim \frac{1}{c^2} \left(\frac{\ell}{\tau} \right)^2 \sim \frac{u^2}{c^2} << 1 \tag{13.15}$$

if we are not dealing with relativistic flows. Therefore a reasonable approximation for (13.11) is

$$\nabla \wedge \mathbf{B} = \mu_0 \mathbf{j} \tag{13.16}$$

Then we can apply a similar argument to considering the ratio of the first and second terms on the right-hand side of (13.7), i.e. the accelerations associated with the $\mathbf{E}$ and $\mathbf{B}$ field respectively. We have just argued that $E/B \sim u$. From (13.16) and (13.10) we have $q/j \sim E/B\epsilon_0\mu_0 \sim u/c^2$. Hence the ratio of $|q\mathbf{E}|$ to $|\mathbf{j} \wedge \mathbf{B}|$ is $\sim (u/c)^2$ and hence $<< 1$ for non-relativistic flows. Thus henceforth we will ignore the Lorentz force associated with the electric field. We note that this argument implies that $q/j \sim (u/c)^2 u^{-1}$ and hence that $q << j/u$. What this means microphysically is that there is a net charge flow (i.e. current) even though the *net charge density* is nearly zero. In order to see how this can come about, consider the equations for $\mathbf{j}$ and q for a fluid with two charged species above equation (13.4): we see that q can be zero while $\mathbf{j}$ is non-zero provided that the two charge species move at slightly different speeds (given the relative masses of the proton and electron, this means that the fluid speed almost coincides with that of the protons, via equation (13.2)).

Since charge neutrality is important for our neglect of the Lorentz force associated with $\mathbf{E}$, we shall consider this issue in a little more detail below, demonstrating that the strength of the electrostatic coupling between protons and electrons is in practice sufficient to ensure charge neutrality to a good approximation.

13.3 Charge neutrality

As an illustration of the strength of electrostatic forces, suppose that there is a positive excess of a small fraction f within a sphere of radius

r in a plasma with a number density n. Electrons on the edge of the sphere experience an acceleration

$$\dot{u} = \frac{eE}{m_e} \sim \frac{4\pi r^3}{3m_e} fn \frac{e^2}{4\pi\epsilon_0 r^2} \qquad (13.17)$$

If we choose $f = 0.01$, $r = 10^{-2}$m (i.e. 1 cm) and $n = 10^{16}\,\mathrm{m}^{-3}$, then $\dot{u} = 10^{15}\,\mathrm{m\,s}^{-2}$, so the region will neutralise on a timescale of $\sim 3 \times 10^{-9}$ s. Indeed the movement is so rapid we would expect some overshoot in the charge imbalance correction, and so develop charge oscillations.

Of course we can investigate this behaviour in more detail, using the usual perturbation approach. We suppose in a region the positive ion number density $n^+ = n_0$ and for the electrons $n_e = n_0 + n_1(r, t)$, with $n_1 << n_0$. Then, using (13.10) we see that the negative charge excess from the electrons implies

$$\nabla \cdot \mathbf{E} = -\frac{1}{\epsilon_0} n_1 e \qquad (13.18)$$

Since $m^+ >> m_e$ we may neglect the contribution of the ion motions – compared with the electrons – to charge transfer (although this does not allow us to neglect the *momentum* associated with ion motion). We treat the electrons as a fluid with velocity field $\mathbf{u}_e(r, t)$ relative to the protons, where u_e is treated as a small quantity, and in the flow equations $\mathbf{u}_e \cdot \nabla\mathbf{u}_e$ is second order in u_e, and so vanishes. The force density is $-n_e e\mathbf{E}$, and the mass density is $n_e m_e$, and so

$$m_e \frac{\partial \mathbf{u}_e}{\partial t} = -e\mathbf{E} \qquad (13.19)$$

Now we use the continuity equation to first order

$$\frac{\partial n_1}{\partial t} + n_0 \nabla \cdot \mathbf{u}_e = 0 \qquad (13.20)$$

$$\Rightarrow \frac{1}{n_0} \frac{\partial^2 n_1}{\partial t^2} - \frac{e}{m_e} \nabla \cdot \mathbf{E} = 0 \qquad (13.21)$$

or

$$\frac{1}{n_0} \frac{\partial^2 n_1}{\partial t^2} + \frac{1}{\epsilon_0} \frac{e^2}{m_e} n_1 = 0 \qquad (13.22)$$

Thus any charge imbalance oscillates with the plasma frequency

$$\omega_P = \left(\frac{1}{\epsilon_0} \frac{n_0 e^2}{m_e} \right)^{\frac{1}{2}} \qquad (13.23)$$

or

$$\nu_P \quad = \frac{\omega_P}{2\pi}$$

$$= \frac{1}{2\pi}\left(\frac{1}{\epsilon_0}\frac{e^2}{m_e}\right)^{\frac{1}{2}}$$

$$\sim 9.0 n_0^{\frac{1}{2}} \text{ Hz} \tag{13.24}$$

In the Earth's ionosphere $n_0 \sim 10^{12}$ m^{-3}, and we find that electromagnetic radiation with $\nu < 10^7$Hz is reflected, because the electrons in the plasma can move to respond rather than allowing the wave to be transmitted.

Associated with the timescale (=1/frequency) there is a length scale $\ell \sim u_e/\omega_P$. If the electron velocity is the thermal velocity $u_e \sim \sqrt{\frac{kT_e}{m_e}}$, then the Debye length

$$\lambda_D = \left(\frac{\epsilon_0 kT_e}{n_0 e^2}\right)^{\frac{1}{2}} \tag{13.25}$$

is an effective shielding length. Thermal motions 'iron out' plasma oscillations on that length scale.

13.4 The induction equation and flux freezing approximation

In order to understand the evolution of magnetic fields in fluid flows, we combine the induction equation (13.13) with the modified form of Ohm's Law (13.9), and make the simplifying assumption that the conductivity σ is constant. Then (by taking the curl of (13.16) and combining with (13.8)) we obtain

$$\mathbf{\nabla} \wedge (\mathbf{\nabla} \wedge \mathbf{B}) = \mu_0 \sigma \left[\mathbf{\nabla} \wedge \mathbf{E} + \mathbf{\nabla} \wedge (\mathbf{u} \wedge \mathbf{B})\right], \tag{13.26}$$

(Here we have assumed that the displacement current term, $\frac{\partial \mathbf{E}}{\partial t}$, is negligible (following the plausibility argument above)). Then we can use (13.12) to eliminate $\mathbf{E}$, and

$$\mathbf{\nabla} \wedge (\mathbf{\nabla} \wedge \mathbf{B}) = -\nabla^2 \mathbf{B} - \mathbf{\nabla}(\mathbf{\nabla} \cdot \mathbf{B}) = -\nabla^2 \mathbf{B} \tag{13.27}$$

since $\mathbf{\nabla} \cdot \mathbf{B} = 0$ (13.9), to obtain

$$\frac{\partial \mathbf{B}}{\partial t} = \mathbf{\nabla} \wedge (\mathbf{u} \wedge \mathbf{B}) + \frac{1}{\mu_0 \sigma} \nabla^2 \mathbf{B}. \tag{13.28}$$

Fig. 13.2.

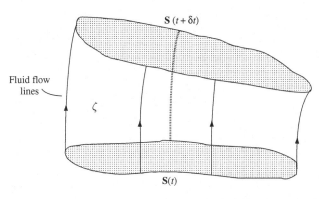

The time rate of change of magnetic field is determined by convection of the field by the fluid ($\mathbf{\nabla} \wedge (\mathbf{u} \wedge \mathbf{B})$) and diffusion through the conductive term $\frac{1}{\mu_0 \sigma} \nabla^2 \mathbf{B}$. We can neglect diffusion if σ is sufficiently large.

For the case of a perfectly conducting fluid ($\sigma = \infty$) we can look at the magnetic flux Φ through a surface $\mathbf{S}(t)$ which is moving with the fluid.

We have

$$\frac{d\Phi}{dt} = \lim_{dt \to 0} \left[\int_{\mathbf{S}(t+dt)} \mathbf{B}(t+dt) \cdot d\mathbf{S} - \int_{\mathbf{S}(t)} \mathbf{B}(t) \cdot d\mathbf{S} \right] \frac{1}{dt}. \qquad (13.29)$$

Since the flux through a closed surface is zero, we have, at time $t + dt$,

$$\int_{\mathbf{S}(t+dt)} \mathbf{B}(t+dt) \cdot d\mathbf{S} + \int_{\zeta} \mathbf{B}(t+dt) \cdot d\zeta - \int_{\mathbf{S}(t)} \mathbf{B}(t+dt) \cdot d\mathbf{S} = 0. \quad (13.30)$$

where ζ is the surface which, together with $\mathbf{S}(t)$ and $\mathbf{S}(t+dt)$, makes up the closed surface containing the paths between time t and $t + dt$ for all the fluid particles passing through $\mathbf{S}$ at time t. But $d\zeta = d\mathbf{x} \wedge \mathbf{u} dt$, where $d\mathbf{x}$ is any line element in the surface ζ, so that

$$\frac{d\Phi}{dt} = \lim_{dt \to 0} \left[\int_{\mathbf{S}(t)} (\mathbf{B}(t+dt) - \mathbf{B}(t)) \cdot d\mathbf{S} - \int_{\zeta} \mathbf{B}(t+dt) \cdot d\mathbf{x} \wedge \mathbf{u} dt \right] \frac{1}{dt}. \qquad (13.31)$$

Thus

$$\frac{d\Phi}{dt} = \int_{\mathbf{S}(t)} \frac{\partial \mathbf{B}}{\partial t} \cdot d\mathbf{S} - \int_{\mathcal{C}} (\mathbf{u} \wedge \mathbf{B}) d\mathbf{x} \qquad (13.32)$$

where $\mathcal{C}$ is any contour in the surface ζ. If $\mathcal{C}$ is taken to be the contour enclosing $\mathbf{S}(t)$, then, using Stokes theorem, we have

$$\frac{d\Phi}{dt} = \int_{\mathbf{S}(t)} \left[\frac{\partial \mathbf{B}}{\partial t} - \mathbf{\nabla} \wedge (\mathbf{u} \wedge \mathbf{B}) \right] \cdot d\mathbf{S}. \qquad (13.33)$$

Therefore from (13.28), we see that the magnetic flux through a surface co-moving with the fluid changes only as a result of the diffusion term $\frac{1}{\mu_0\sigma}\nabla^2\mathbf{B}$. If we are in the high conductivity limit. so that the diffusion term is negligible, we have

$$\frac{d\Phi}{dt} = 0. \tag{13.34}$$

(The derivation for the conservation of vorticity flux through a surface moving with the fluid element is exactly the same as this – see Chapter 9.)

Therefore, under these circumstances, we have shown that the fluid and the magnetic field move exactly together, in the sense that a magnetic field line consists of the same fluid particles at all times. (This can readily be shown by applying the above result to an infinitesimal surface element). We say that a magnetic field is *frozen in* to a perfectly conducting fluid.

As a further consequence of infinite conductivity, and finite currents, we must have

$$\mathbf{E} + \mathbf{u} \wedge \mathbf{B} = 0 \tag{13.35}$$

and so (taking this $\cdot \mathbf{B}$)

$$\mathbf{E} \cdot \mathbf{B} = 0 \tag{13.36}$$

i.e. electric and magnetic fields are perpendicular.

13.5 The dynamical effects of magnetic fields

To get an idea of what the presence of a magnetic field does, it is worth looking at the magnetic forces on a plasma. We know the magnetic force density is

$$\mathbf{f}_{mag} = \mathbf{j} \wedge \mathbf{B} \tag{13.37}$$

and that

$$\nabla \wedge \mathbf{B} = \mu_0 \mathbf{j} \tag{13.38}$$

from (13.16). Eliminating $\mathbf{j}$ yields

$$\mathbf{f}_{mag} = \frac{1}{\mu_0}(\nabla \wedge \mathbf{B}) \wedge \mathbf{B} \tag{13.39}$$

and this is equivalent to

$$\mathbf{f}_{\text{mag}} = \frac{1}{\mu_0}\left[-\nabla\left(\frac{B^2}{2}\right) + (\mathbf{B}\cdot\nabla)\mathbf{B}\right].$$ (13.40)

Now comparing this with the pressure term in the Euler equations, or the general MHD equations, we see that the first term on the RHS of (13.40) behaves like a hydrostatic pressure of magnitude

$$p_{\text{mag}} = \frac{B^2}{2\mu_0}.$$ (13.41)

The final term in (13.40) is better looked at by considering special cases. If in a cylindrical coordinate system (R, θ, z) one chooses $\mathbf{B} = (0, B_0, 0)$, for example, and writes out the resultant terms, then it becomes evident that the second term on the RHS of (13.40) is equivalent to a tension per unit area of magnitude B^2/μ_0 along the field lines.

Just looking at the order of magnitude of the terms in the Euler equation shows that the relative importance of the ram pressure, gas pressure and magnetic pressure terms is given by the relative sizes of $\rho u^2/2$, ρc_s^2 and $B^2/2\mu_0$, where c_s is the sound speed in the gas. If the magnetic term is small, then it is generally a sign that magnetic fields have a small effect on the fluid motions, and so to a good approximation can be omitted.

The equality of kinetic and magnetic energy densities defines a velocity

$$v_{\text{A}} = \left(\frac{B^2}{\rho\mu_0}\right)^{\frac{1}{2}},$$ (13.42)

the Alfvén velocity. We will shortly see that this velocity is important in defining the propagation speed of disturbances in magnetised plasmas.

13.6 Summary

The equations of ideal magneto-hydrodynamics (in the absence of cooling and viscosity) are described by:

$$\frac{\partial\rho}{\partial t} + \nabla\cdot(\rho\mathbf{u}) = 0$$

$$\rho\left(\frac{\partial\mathbf{u}}{\partial t} + \mathbf{u}\cdot\nabla\mathbf{u}\right) = \mathbf{j}\wedge\mathbf{B} - \nabla p$$

$$p = K\rho^{\frac{5}{3}}$$ (13.43)

$$\nabla \wedge (\mathbf{u} \wedge \mathbf{B}) = \frac{\partial \mathbf{B}}{\partial t}$$

$$\nabla \wedge \mathbf{B} = \mu_0 \mathbf{j}$$

$$\nabla \cdot \mathbf{B} = 0 \tag{13.44}$$

Note that by writing the relationship between pressure and density in this way we are assuming that the fields do no work on the system: this is evidently true in the case of an infinitely conducting medium since in this case $\mathbf{E}$ is perpendicular to $\mathbf{u}$.

It is perhaps worth recapitulating the main assumptions behind these equations:

(i) neglect of the displacement current (second term in (13.11)) since the flow is highly non-relativistic

(ii) assumption of infinite conductivity, which implies flux freezing and no work done on the fluid by the fields

(iii) negligible charge separation (which allows neglect of the Lorentz force associated with $\mathbf{E}$.)

13.7 Waves in Plasmas

We have the equation set (13.44) and want to examine the nature of wave solutions. As usual, we take the simplest situation possible, and consider a uniform density ideal fluid at rest, which is threaded by a uniform magnetic field $\mathbf{B}_0$. Then consider a small disturbance to the equilibrium situation, with the small displacements denoted by subscript 1. We could write down the standard solution and set up a dispersion relation straight away, but it is a little more useful to try to simplify the $\wedge$ terms in the first order equations before doing this.

The continuity equation becomes

$$\frac{\partial \rho_1}{\partial t} + \rho_0 \nabla \cdot \mathbf{u}_1 = 0. \tag{13.45}$$

The momentum equation is

$$\rho_0 \frac{\partial \mathbf{u}_1}{\partial t} = -\nabla p_1 - \frac{1}{\mu_0} \mathbf{B}_0 \wedge (\nabla \wedge \mathbf{B}_1), \tag{13.46}$$

since the initial field is uniform, so $\nabla \wedge \mathbf{B}_0 = 0$. The $p(\rho)$ relation yields

$$p_1/p_0 = \gamma \rho_1/\rho_0, \tag{13.47}$$

so (13.46) becomes

$$\rho_0 \frac{\partial \mathbf{u}_1}{\partial t} + c_s^2 \nabla \rho_1 + \frac{1}{\mu_0} \mathbf{B}_0 \wedge (\nabla \wedge \mathbf{B}_1) = 0. \tag{13.48}$$

The **B** equation is

$$\frac{\partial \mathbf{B}}{\partial t} = \nabla \wedge (\mathbf{u} \wedge \mathbf{B}), \tag{13.49}$$

and this to first order in the perturbed quantities becomes

$$\frac{\partial \mathbf{B}_1}{\partial t} = \nabla \wedge (\mathbf{u}_1 \wedge \mathbf{B}_0). \tag{13.50}$$

Now if we take $\frac{\partial}{\partial t}$ of equation (13.48)$\div \rho_0$ this gives us

$$\frac{\partial^2 \mathbf{u}_1}{\partial t^2} + \frac{c_s^2}{\rho_0} \nabla \frac{\partial \rho_1}{\partial t} + \frac{\mathbf{B}_0}{\mu_0 \rho_0} \wedge (\nabla \wedge \frac{\partial \mathbf{B}_1}{\partial t}) = 0. \tag{13.51}$$

Substituting for $\frac{\partial \rho_1}{\partial t}$ from (13.46), and $\frac{\partial \mathbf{B}_1}{\partial t}$ from (13.50), gives us

$$\frac{\partial^2 \mathbf{u}_1}{\partial t^2} - c_s^2 \nabla(\nabla \cdot \mathbf{u}_1) + \mathbf{v}_A \wedge \nabla \wedge [\nabla \wedge (\mathbf{u}_1 \wedge \mathbf{v}_A)] = 0, \tag{13.52}$$

where we have introduced a vectorial *Alfvén velocity*

$$\mathbf{v}_A = \frac{\mathbf{B}_0}{\sqrt{\mu_0 \rho_0}}. \tag{13.53}$$

This wave equation for $\mathbf{u}_1$ is rather involved, but it allows simple solutions for waves parallel or perpendicular to the magnetic field direction. We use the usual form and set

$$\mathbf{u}_1 = \tilde{\mathbf{u}}_1 \exp i(\mathbf{k} \cdot \mathbf{x} - \omega t). \tag{13.54}$$

We can write $\mathbf{v}_A \wedge \nabla \wedge [\nabla \wedge (\mathbf{u}_1 \wedge \mathbf{v}_A)]$ in component notation as

$$\epsilon_{jlm} v_l \epsilon_{mnp} \partial_n \epsilon_{pqr} \partial_q \epsilon_{rst} u_s v_t, \tag{13.55}$$

where we use u_i for the components of $\mathbf{u}_1$, and v_i for the components of $\mathbf{v}_A$ to save a profusion of indices. Then we use the permutation symbol relation $\epsilon_{ijk} \epsilon_{klm} = (\delta_{il} \delta_{jm} - \delta_{jl} \delta_{im})$ to 'simplify' this expression. At the same time, below, we substitute the exponential terms where the derivatives apply.

$$\mathbf{v}_A \wedge \nabla \wedge [\nabla \wedge (\mathbf{u}_1 \wedge \mathbf{v}_A)] \tag{13.56}$$

$$= \epsilon_{jlm} v_l \epsilon_{mnp} \partial_n \epsilon_{pqr} \partial_q \epsilon_{rst} u_s v_t$$

$$= (\delta_{jn} \delta_{lp} - \delta_{ln} \delta_{jp}) v_l \partial_n (\delta_{ps} \delta_{qt} - \delta_{qs} \delta_{pt}) \partial_q u_s v_t$$

$$= (v_p \partial_j - v_l \partial_l \delta_{jp})(\partial_q u_p v_q - \partial_q u_q v_p)$$

$$= v_p v_q \partial_j \partial_q u_p - v_p v_p \partial_j \partial_q u_q - v_q v_l \partial_l \partial_q u_j + v_l v_j \partial_l \partial_q u_q$$

Now set

$$u_i = \tilde{u}_i e^{i(\mathbf{k} \cdot \mathbf{x} - \omega t)}, \tag{13.57}$$

so

$$\partial_j u_i = \tilde{u}_i e^{i(\mathbf{k} \cdot \mathbf{x} - \omega t)} i k_j \tag{13.58}$$

and then

$$v_p v_q \partial_j \partial_q u_p - v_p v_p \partial_j \partial_q u_q - v_q v_l \partial_l \partial_q u_j + v_l v_j \partial_l \partial_q u_q \tag{13.59}$$
$$= (v_p v_q \tilde{u}_p i k_j i k_q - v_p v_p \tilde{u}_q i k_j i k_q - v_q v_l \tilde{u}_j i k_l i k_q + v_l v_j \tilde{u}_q i k_l i k_q) \times$$
$$e^{i(\mathbf{k} \cdot \mathbf{x} - \omega t)}$$
$$= [-(\mathbf{v}_A \cdot \tilde{\mathbf{u}})(\mathbf{v}_A \cdot \mathbf{k})\mathbf{k} + (\mathbf{v}_A \cdot \mathbf{v}_A)(\mathbf{k} \cdot \tilde{\mathbf{u}})\mathbf{k} + (\mathbf{v}_A \cdot \mathbf{k})(\mathbf{v}_A \cdot \mathbf{k})\tilde{\mathbf{u}}$$
$$-(\mathbf{v}_A \cdot \mathbf{k})(\mathbf{k} \cdot \tilde{\mathbf{u}})\mathbf{v}_A] e^{i(\mathbf{k} \cdot \mathbf{x} - \omega t)}$$

Also

$$c_s^2 \mathbf{\nabla}(\mathbf{\nabla} \cdot \mathbf{u}_1) = -c_s^2 \partial_j \partial_k \mathbf{u}_k = -c_s^2 (\mathbf{k} \cdot \tilde{\mathbf{u}}) \mathbf{k} e^{i(\mathbf{k} \cdot \mathbf{x} - \omega t)} \tag{13.60}$$

and

$$\frac{\partial^2 \mathbf{u}_1}{\partial t^2} = -\omega^2 \tilde{\mathbf{u}} e^{i(\mathbf{k} \cdot \mathbf{x} - \omega t)} \tag{13.61}$$

Then (13.52) becomes

$$-\omega^2 \mathbf{u}_1 + (c_s^2 + v_A^2)(\mathbf{k} \cdot \mathbf{u}_1)\mathbf{k} + (\mathbf{v}_A \cdot \mathbf{k})[(\mathbf{v}_A \cdot \mathbf{k})\mathbf{u}_1 \tag{13.62}$$
$$-(\mathbf{v}_A \cdot \mathbf{u}_1)\mathbf{k} - (\mathbf{k} \cdot \mathbf{u}_1)\mathbf{v}_A] = 0.$$

If $\mathbf{k}$ is perpendicular to $\mathbf{v}_A$ the last term vanishes. (Recall that $\mathbf{k}$ is perpendicular to lines of constant phase, whereas $\mathbf{v}_A$ is parallel to the unperturbed field) The solution is then a longitudinal magnetosonic wave with phase velocity $\sqrt{c_s^2 + v_A^2}$, so it propagates with a velocity that depends on the sum of the hydrostatic and magnetic pressures (to within factors of order unity). The magnetic part of this term comes from the magnetic pressure term in equation (13.40). We may easily envisage what is happening in this wave: longitudinal disturbances result in successive rarefactions and compressions, just as in an ordinary sound wave. In this case, however, the magnetic field lines are bunched together in the compressions since the field is 'frozen' to the fluid and this means that there is an additional (magnetic) pressure resisting the compression. Consequently, the wave speed has contributions from both the thermal and magnetic pressure.

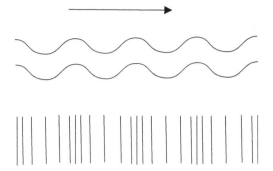

On the other hand, if $\mathbf{k}$ is parallel to $\mathbf{v}_A$, (13.62) becomes

$$(k^2 v_A^2 - \omega^2)\mathbf{u}_1 + \left(\frac{c_s^2}{v_A^2} - 1\right) k^2 (\mathbf{v}_A \cdot \mathbf{u}_1)\mathbf{v}_A = 0. \tag{13.63}$$

In this case there are two types of wave motion. There is an ordinary longitudinal wave, with $\mathbf{u}_1$ parallel to $\mathbf{k}$ and $\mathbf{v}_A$, which travels at the sound speed c_s. In this case, the compressions and rarefactions leave the field undisturbed; fluid in such motion experiences no hydromagnetic forces and so the wave speed 'knows' only about the thermal pressure c_s. There is also a transverse wave, $(\mathbf{v}_A \cdot \mathbf{u}_1 = 0)$ with a phase velocity equal to the Alfvén speed v_A. This wave is a purely magneto-hydrodynamic wave, which depends effectively on the tension in the magnetic field lines and the inertia of the material which moves with the field, since the field is 'frozen in'. We may envisage this sort of wave as being analogous to a transverse wave on a string; the restoring force of displaced fluid elements is provided by magnetic tension, just as it is provided by the tension of a material string. Evidently, such transverse solutions *require* a magnetic field, since thermal pressure which acts normally on any surface in the fluid cannot generate transverse stresses.

The magnetic fields of these different waves can be found from equation (13.50). For $\mathbf{k}$ perpendicular to $\mathbf{B}_0$ we have

$$\mathbf{B}_1 = \frac{k}{\omega} u_1 \mathbf{B}_0, \tag{13.64}$$

and in the case where $\mathbf{k}$ is parallel to $\mathbf{B}_0$

$$\mathbf{B}_1 = 0 \quad \text{for longitudinal waves} \tag{13.65}$$

$$\mathbf{B}_1 = -\frac{k}{\omega} u_1 \mathbf{B}_0 \quad \text{for transverse waves.} \tag{13.66}$$

In astrophysical situations, the Alfvén speed can be comparable with, or larger than, the thermal sound speed. In the solar photosphere, for example, the density is about 10^{-4} kg m^{-3} (which is about 6×10^{22} hydrogen atoms m^{-3}), so that $v_A \approx 10^5 B$ ms^{-1} (B in Tesla). Solar magnetic fields are typically a few$\times 10^{-4}$T at the surface, with significantly higher values in (~ 0.3T) in sunspots. The effective temperature at the solar surface is 5770K, so the sound speed a little under 10^4 m s^{-1}.

Another environment where Alfvén waves play an important role is within giant molecular cloud complexes, where the kinetic and magnetic energy densities are comparable, and exceed the thermal energy of the very cold gas in these clouds by more than an order of magnitude. This means that shocks may be 'softened' in this environment: i.e. fluid elements may collide at speeds that are supersonic but sub-Alfvénic. As a result, whereas a strong shock would occur in an unmagnetised medium, this may be avoided in the presence of the magnetic field, since an Alfven wave may be able to carry 'news' of the impending collision fast enough to 'soften the blow'. It was believed at one time that this effect would be enough to avoid strongly inelastic collisions within molecular clouds and that this would therefore extend the timescale on which such clouds would dissipate their internal motions and collapse. More realistic multi-dimensional simulations of magnetohydrodynamical turbulence in these clouds have shown that actually the lifetime of the complex is *not* significantly extended in this way, since if the fluid motions are turbulent there will always be situations where fluid elements will be colliding *along* field lines and where Alfven waves will be of no avail. Nevertheless, much of the structure of the dense interstellar medium can only be understood in terms of magnetised shocks: since the field structure determines the compression and heating in the shock, the abundances of certain temperature sensitive molecules in shocked regions provides a good diagnostic of the strength and local topology of magnetic fields in these regions.

13.8 The Rayleigh-Taylor Instability revisited

Since the motions of a plasma and a magnetic field are so closely tied together, one might expect that if the magnetic term dominates it might be possible to stablise a stratified fluid in which a higher density fluid is above a lower density one in a gravitational field. In the absence of any magnetic field this configuration is Rayleigh-Taylor unstable, as was shown in Chapter 10. In practice, astrophysical environments where the Rayleigh Taylor instability is important (i.e. the convective regions of stars or where a supernova blast wave is decelerated by interaction with the insterstellar medium) may also contain dynamically significant magnetic fields. We therefore need to examine how the instability is modified by such fields.

We suppose we have two incompressible fluids at rest, one above the other in a uniform gravitational field. Now we apply a uniform magnetic field with field lines parallel to the interface between the fluids, and ask if it helps stabilise them when the upper one is of greater density than the lower one.

Fig. 13.4.

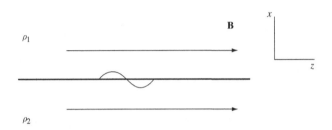

We take each fluid to be incompressible, and suppose the displacement in the fluid

$$\vec{\xi} = \vec{\xi}(x)e^{i(kz-\omega t)} \tag{13.67}$$

and similarly for the other variables (i.e. we assume that wavelike propagation is parallel to the interface and with an amplitude that depends on distance from the interface). Note also that $\vec{\xi}$ is a vector, i.e. it can have components in each of x and y and z. The equations are then

$$\frac{\partial \rho}{\partial t} + \nabla \cdot (\rho \mathbf{u}) = 0 \tag{13.68}$$

$$\rho \left(\frac{\partial \mathbf{u}}{\partial t} + \mathbf{u} \cdot \nabla \mathbf{u} \right) = \mathbf{j} \wedge \mathbf{B} - \nabla p + \rho \mathbf{g} \tag{13.69}$$

where $\mathbf{g} = (-g, 0, 0)$ and in the unperturbed state, $\mathbf{B} = (0, 0, B^0)$,

$$\frac{\partial \mathbf{B}}{\partial t} = \nabla \wedge (\mathbf{u} \wedge \mathbf{B}) \tag{13.70}$$

$$\nabla \wedge \mathbf{B} = \mu_0 \mathbf{j}. \tag{13.71}$$

Then

$$\mathbf{u} = \frac{\partial \vec{\xi}}{\partial t} = -i\omega \vec{\xi}(x)e^{i(kz-\omega t)} = -i\omega \vec{\xi} \tag{13.72}$$

Equation (13.69) becomes

$$-\rho\omega^2\boldsymbol{\xi} = -\hat{\mathbf{e}}_x\left(\frac{dp}{dx}+\rho g\right) - \hat{\mathbf{e}}_z\frac{dp}{dz}+\mathbf{j}\wedge\mathbf{B}^0 \tag{13.73}$$

where we use $\mathbf{B}^0$ rather than $\mathbf{B}$ because $\mathbf{j}=0$ in the unperturbed case, and so in general it will be a first order quantity.

In component terms this becomes

$$-\rho\omega^2\xi_x = -\frac{dp}{dx}-\rho g+(\mathbf{j}\wedge\mathbf{B}^0)_x \tag{13.74}$$

and

$$-\rho\omega^2\xi_z = -ikp^1 \tag{13.75}$$

where p^1 is the perturbed pressure value. Within the individual fluids the incompressibility condition implies $\nabla\cdot\mathbf{u}=0$, so using the relationship between velocity and $\boldsymbol{\xi}$ from (13.72), gives

$$\frac{d\xi_x}{dx}+ik\xi_z = 0 \tag{13.76}$$

The magnetic field term comes from (13.70). Using the form for the perturbed value $\mathbf{B}^1 = \mathbf{B}^1(x)e^{i(kz-\omega t)}$ and the velocity from (13.72) again, we have

$$-i\omega\mathbf{B}^1 = \nabla\wedge(-i\omega\boldsymbol{\xi}\wedge\mathbf{B}^0) \tag{13.77}$$

i.e.

$$\mathbf{B}^1 = \nabla\wedge(\boldsymbol{\xi}\wedge\mathbf{B}^0) \tag{13.78}$$

Since the only non-zero component of $\mathbf{B}^0$ is the z-component, we find from (13.78) that

$$B_x^1 = B^0 ik\xi_x \quad\text{and}\quad B_z^1 = -B^0\frac{d\xi_x}{dx} \tag{13.79}$$

We can then use Ampère's law (equation (13.71)) to obtain an expression for $\mathbf{j}$. The x- and z-terms are zero and the only non-zero current term is

$$j_y = B^0\left(\frac{d^2\xi_x}{dx^2}-k^2\xi_x\right). \tag{13.80}$$

The pressure-gravity term in (13.69) can be simplified, since in the equilibrium state $\frac{dp}{dx} = -g\rho$, so all we have to evaluate is $\frac{dp^1}{dx}$. We know that

$$p^1 = \frac{\rho\omega^2}{ik}\xi_z \tag{13.81}$$

from (13.75), so, using (13.76),

$$p^1 = \rho\left(\frac{\omega^2}{k^2}\right)\frac{d\xi_x}{dx} \tag{13.82}$$

and

$$\frac{dp^1}{dx} = \frac{d}{dx}\left[\rho\left(\frac{\omega^2}{k^2}\right)\frac{d\xi_x}{dx}\right]. \tag{13.83}$$

The final term we have to consider is the perturbed density. The fluids are incompressible so there is no density variation within each component, but we do have to consider the density dependence near the boundary in the Eulerian formulation. The mass conservation equation in linearised form (or the Lagrangian one transformed) gives us

$$-i\omega\rho^1 - i\omega\boldsymbol{\xi} \cdot \nabla\rho_0 = 0 \tag{13.84}$$

or

$$\rho^1 = -\xi_x\frac{d\rho_0}{dx}. \tag{13.85}$$

Using all these we are now in a position to write down the momentum equation in terms of ξ_x alone. It is

$$\frac{d}{dx}\left(\rho_0\frac{d\xi_x}{dx}\right) - k^2\rho_0\xi_x - \left(\frac{k}{\omega}\right)^2 g\frac{d\rho_0}{dx}\xi_x - \frac{1}{\mu_0}\left(\frac{kB^0}{\omega}\right)^2\left[\frac{d^2\xi_x}{dx^2} - k^2\xi_x\right] = 0. \tag{13.86}$$

As in the case where there is no magnetic field, this equation needs to be integrated with respect to x across the interface separating the two fluids, whilst applying the boundary conditions that the fluid displacement is zero as $x \to \pm\infty$ and that ξ_x is continuous across the boundary at $x = 0$. This is similar to the magnetic field-free case we dealt with earlier. The values of ω which permit these boundary conditions are given by the dispersion relation

$$\omega^2 = kg\left(\frac{\rho_2 - \rho_1}{\rho_2 + \rho_1}\right) + \frac{2}{\mu_0}\left(\frac{k^2 B_0^2}{\rho_2 + \rho_1}\right) \tag{13.87}$$

The first term on the RHS can be positive or negative, depending on the relative densities of the two fluids. As we saw, in the absence of a magnetic field, $\omega^2 < 0$ if $\rho_2 < \rho_1$, so the system is unstable. The second term on the RHS is always positive, so the presence of the magnetic field does help to stabilise the configuration. This is because work is expended by the fluid in bending the field lines. Because the field line bending term is proportional to k^2, the stabilising effect is stronger for short wavelength modes. We may write a necessary criterion for the critical wavelength of the modes

$$k > k_{\text{crit}} = \frac{g\mu_0}{2B_0^2}(\rho_1 - \rho_2).$$

(13.88)

However, the situation set up here is rather a special one, in that the oscillatory term is in the same direction as the magnetic field. If the perturbation is at some arbitrary angle with respect to the field then the dispersion relation becomes

$$\omega^2 = kg\left(\frac{\rho_2 - \rho_1}{\rho_2 + \rho_1}\right) + \frac{2}{\mu_0}\frac{(\mathbf{k} \cdot \mathbf{B}_0)^2}{\rho_2 + \rho_1}.$$

(13.89)

This is the same as before when $\mathbf{k}$ is parallel to $\mathbf{B}$, but when $\mathbf{k}$ is perpendicular to $\mathbf{B}$ there is no stabilising effect at all. Thus some

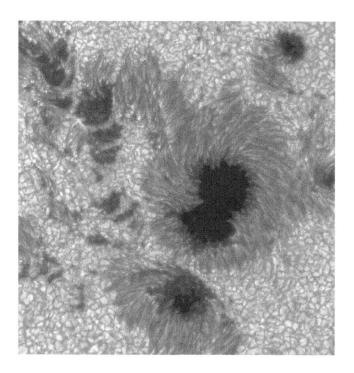

Fig. 13.5. A sunspot on the surface of the sun, taken on 15 July, 2002. The large one in the centre of the picture has a long axis of ~ 15000 km. (*Royal Swedish Academy of Sciences, Institute for Solar Physics*)

growth modes are suppressed by the magnetic field, but not all. In practice, this means that there are no situations where the Rayleigh Taylor instability is completely stabilised by magnetic fields but that in the case of strong fields, only perturbations propagating close to perpendicular to the interface will be able to grow in the linear regime. Convection is thus modified in the presence of strong fields. It is this effect, along with the lower gas pressure in magnetic regions required to balance the gas pressure in outside regions of the solar photosphere, which accounts for the low temperature (hence dark appearance) of sunspots.

Appendix
Equations in curvilinear coordinates

When it comes to applying the fluid equations we generally wish to use a coordinate system in which we minimize the number of equations by using symmetries inherent in the problem to set some derivatives in suitable coordinates to zero. The obvious such coordinate systems are spherical polars, which will apply in the simplest stellar situations for example, and cylindrical polars in those cases where there is some axial symmetry, usually for rotating systems. Here we provide expressions for the fluid equations in these coordinate systems, and give expressions for the various vector differential operators.

A.1 Tensor approach

Those familiar with tensors can reformulate the fluid equations directly, and then use the tensor formalism to obtain these equations in any coordinate system. Those who are not familar with tensor methods should skip to the relevant formulae in Section A.3 or A.4.

All we have to do with each of the fluid equations is note that the coordinates and their time derivatives are contravariant vectors, replace Cartesian coordinate derivatives with covariant derivatives, and use the metric tensor g_{ij}, where the line element ds is given by

$$ds^2 = g_{ij}\, dx^i\, dx^j, \tag{A.1}$$

to raise or lower indices as appropriate.

The Eulerian continuity equation (2.4) becomes

$$\frac{\partial \rho}{\partial t} + (\rho v^i)_{;i} = 0, \tag{A.2}$$

where there is the usual implicit sum over the spatial components with axis labels x^i for $i = 1$ to 3, and the contravariant velocity vector v^i

is the time derivative of the spatial coordinate $v^i = \frac{\partial x^i}{\partial t}$. The semicolon indicates the covariant derivative, so for any contravariant vector T^i

$$T^i_{;j} = \frac{\partial T^i}{\partial x^j} + \Gamma^i_{kj} T^k, \tag{A.3}$$

where the Γ^i_{kj} are Christoffel symbols of the second kind, and for a scalar quantity Φ

$$\Phi_{;j} = \frac{\partial \Phi}{\partial x^j}. \tag{A.4}$$

The momentum equation (4.2) is

$$\frac{\partial v^i}{\partial t} + v^j v^i_{;j} = -g^{ij} \left(\frac{1}{\rho} p_{;j} + \Psi_{;j} \right), \tag{A.5}$$

where p is the pressure and Ψ the gravitational potential.

Poisson's equation for the gravitational potential (3.18) becomes

$$\left(g^{ij} \Psi_{;j} \right)_{;i} = 4\pi G \rho, \tag{A.6}$$

and the energy equation (4.32) is

$$\frac{\partial E}{\partial t} + \left[(E + p) v^i \right]_{;i} = -\rho \dot{Q}_{\text{cool}} + \rho \frac{\partial \Psi}{\partial t}. \tag{A.7}$$

These equations apply in any coordinate system we care to choose, not only the usual Cartesian, cylindrical polar and spherical polar coordinate systems. The metric need not even be diagonal, but if it is then there are some general forms for the vector operators which can be used.

A.2 Div, grad and curl in orthogonal curvilinear coordinates

For a diagonal metric we can express the differential operators as derivatives on components with the metric coefficients. Here we state the results. The metric is taken to be

$$ds^2 = h_1^2 \, dq_1^2 + h_2^2 \, dq_2^2 + h_3^2 \, dq_3^2 \tag{A.8}$$

and $\hat{\mathbf{e}}_i$ denotes the unit vector along the q_i axis. Then (forgetting the summation convention used above for the moment)

$$\nabla \Psi = \sum_i \frac{1}{h_i} \frac{\partial \Psi}{\partial q_i} \mathbf{e}_i, \tag{A.9}$$

$$\nabla \cdot \mathbf{A} = \frac{1}{h_1 h_2 h_3} \sum_{\substack{i,j,k \\ \text{cyclic}}} \frac{\partial}{\partial q_i} \left(h_j h_k A_i \right), \tag{A.10}$$

where the notation 'i, j, k cyclic' means sum over i with j and k chosen cyclically depending on i. So for $i = 1$ set $j = 2$ and $k = 3$, for $i = 2$ $j = 3$ and $k = 1$, and for $i = 3$ $j = 1$ and $k = 2$.

For completeness

$$\nabla \wedge \mathbf{A} = \sum_{\substack{i,j,k \\ \text{cyclic}}} \frac{1}{h_j h_k} \left[\frac{\partial}{\partial q_j}(h_k A_k) - \frac{\partial}{\partial q_k}(h_j A_j) \right] \hat{\mathbf{e}}_i. \qquad (A.11)$$

Others can then be formed by combining these, so e.g.

$$\nabla^2 \Psi = \frac{1}{h_1 h_2 h_3} \sum_{\substack{i,j,k \\ \text{cyclic}}} \frac{\partial}{\partial q_i} \left(\frac{h_j h_k}{h_i} \frac{\partial \Psi}{\partial q_i} \right). \qquad (A.12)$$

Also

$$(\mathbf{B} \cdot \nabla)\mathbf{A} = \sum_{\substack{i,j,k \\ \text{cyclic}}} \left[(\mathbf{B} \cdot \nabla)A_i + \frac{A_j}{h_i h_j} \left(B_i \frac{\partial h_i}{\partial q_j} - B_j \frac{\partial h_j}{\partial q_i} \right) \right.$$

$$\left. + \frac{A_k}{h_i h_k} \left(B_i \frac{\partial h_i}{\partial q_k} - B_k \frac{\partial h_k}{\partial q_i} \right) \right] \hat{\mathbf{e}}_i. \qquad (A.13)$$

A.3 Fluid equations in spherical polar coordinates

For spherical polars (r, θ, ϕ) the spatial metric is

$$ds^2 = dr^2 + r^2 d\theta^2 + r^2 \sin^2 \theta \, d\phi^2. \qquad (A.14)$$

The metric g_{ij} is diagonal, so we can evaluate the Γ^i_{jk} quite quickly using the expression below, which apply in these circumstances for $i \neq j \neq k$ and with no summation over repeated indices:

$$\Gamma^i_{jk} = 0, \quad \Gamma^i_{jj} = -\frac{1}{2g_{ii}} \frac{\partial g_{jj}}{\partial x^i}, \quad \Gamma^i_{ji} = \Gamma^i_{ij} = \frac{1}{2g_{ii}} \frac{\partial g_{ii}}{\partial x^j}, \quad \Gamma^i_{ii} = \frac{1}{2g_{ii}} \frac{\partial g_{ii}}{\partial x^i}.$$

The non-zero Christoffel symbols are then

$$\Gamma^1_{22} = -r \qquad \Gamma^1_{33} = -r \sin^2 \theta \quad \Gamma^2_{33} = -r \sin \theta \cos \theta$$

$$\Gamma^2_{12} = \Gamma^2_{21} = \frac{1}{r} \quad \Gamma^3_{13} = \Gamma^3_{31} = \frac{1}{r} \quad \Gamma^3_{23} = \Gamma^3_{32} = \cot \theta$$

so the continuity equation becomes

$$\frac{\partial \rho}{\partial t} + \frac{\partial}{\partial r}(\rho v^r) + \frac{\partial}{\partial \theta}(\rho v^\theta) + \frac{\partial}{\partial \phi}(\rho v^\phi) + \frac{2}{r}(\rho v^r) + \cot \theta \, (\rho v^\theta) = 0. \quad (A.15)$$

We now have the slight complication that we have defined the velocity components $(u_r, u_\theta, u_\phi) = (v_r, v_\theta r, v_\phi r \sin\theta)$ so we have

$$\frac{\partial \rho}{\partial t} + \frac{\partial}{\partial r}(\rho u^r) + \frac{1}{r}\frac{\partial}{\partial \theta}(\rho u^\theta) + \frac{1}{r\sin\theta}\frac{\partial}{\partial \phi}(\rho u^\phi) + \frac{2}{r}(\rho u^r) + \frac{1}{r}\cot\theta\,(\rho u^\theta) = 0.$$
(A.16)

This is equivalent to

$$\frac{\partial \rho}{\partial t} + \frac{1}{r^2}\frac{\partial}{\partial r}(r^2 \rho u^r) + \frac{1}{r\sin\theta}\frac{\partial}{\partial \theta}(\rho \sin\theta\, u^\theta) + \frac{1}{r\sin\theta}\frac{\partial}{\partial \phi}(\rho u^\phi) = 0, \quad \text{(A.17)}$$

which is also the result we would have obtained by applying equation (A.10) with $\mathbf{A} = \rho\mathbf{u}$.

The other fluid equations are dealt with in a similar way. The contravariant metric tensor has non-zero components $g^{11} = 1$, $g^{22} = 1/r^2$ and $g^{33} = 1/r^2 \sin^2\theta$, so after some manipulation the momentum equation becomes, in coordinate terms,

$$\frac{\partial u^r}{\partial t} + u^r\frac{\partial u^r}{\partial r} + \frac{u^\theta}{r}\left(\frac{\partial u^r}{\partial \theta} - u^\theta\right) + \frac{u^\phi}{r\sin\theta}\left(\frac{\partial u^r}{\partial \phi} - u^\phi \sin\theta\right) = -\frac{1}{\rho}\frac{\partial p}{\partial r} - \frac{\partial \Psi}{\partial r},$$
(A.18)

$$\frac{\partial(u^\theta/r)}{\partial t} + \frac{u^r}{r}\frac{\partial u^\theta}{\partial r} + \frac{u^\theta}{r^2}\frac{\partial u^\theta}{\partial \theta} + \frac{u^\theta u^r}{r^2}$$
$$+ \frac{1}{r^2\sin\theta}\left(u^\phi\frac{\partial u^\theta}{\partial \phi} - (u^\phi)^2\cos\theta\right) = -\frac{1}{\rho}\frac{\partial p}{\partial \theta} - \frac{\partial \Psi}{\partial \theta},$$
(A.19)

$$\frac{\partial(u^\phi/r\sin\theta)}{\partial t} + \frac{1}{r^2\sin^2\theta}\left(u^r\frac{\partial u^\phi}{\partial r}r\sin\theta + u^\theta\frac{\partial u^\phi}{\partial \theta}\sin\theta\right.$$
$$\left.+ u^\phi\frac{\partial u^\phi}{\partial \phi} + u^\phi u^r\sin\theta + u^\phi u^\theta\cos\theta\right) = -\frac{1}{\rho}\frac{\partial p}{\partial \phi} - \frac{\partial \Psi}{\partial \phi}.$$
(A.20)

Poisson's equation is

$$\frac{1}{r^2}\frac{\partial}{\partial r}\left(r^2\frac{\partial \Psi}{\partial r}\right) + \frac{1}{r^2\sin\theta}\frac{\partial}{\partial \theta}\left(\sin\theta\frac{\partial \Psi}{\partial \theta}\right) + \frac{1}{r^2\sin^2\theta}\frac{\partial^2 \Psi}{\partial \phi^2} = 4\pi G\rho. \quad \text{(A.21)}$$

The energy equation becomes

$$\frac{\partial E}{\partial t} + \frac{1}{r^2}\frac{\partial}{\partial r}\left[r^2(E+p)u^r\right] + \frac{1}{r\sin\theta}\frac{\partial}{\partial \theta}\left[(E+p)u^\theta\sin\theta\right]$$
$$+ \frac{1}{r\sin\theta}\frac{\partial}{\partial \phi}\left[(E+p)u^\phi\right] = -\rho\dot{Q}_{\text{cool}} + \rho\frac{\partial \Psi}{\partial t}.$$
(A.22)

None of these looks very attractive before a spherical symmetry condition is imposed. Then all the $\frac{\partial}{\partial \theta}$ and $\frac{\partial}{\partial \phi}$ terms become zero and, if we consider only radial flows so we can assume that $u^\theta = u^\phi = 0$, we are left with

$$\frac{\partial \rho}{\partial t} + \frac{1}{r^2}\frac{\partial}{\partial r}(r^2 \rho u^r) = 0, \tag{A.23}$$

$$\frac{\partial u^r}{\partial t} + u^r \frac{\partial u^r}{\partial r} = -\frac{1}{\rho}\frac{\partial p}{\partial r} - \frac{\partial \Psi}{\partial r}, \tag{A.24}$$

$$\frac{1}{r^2}\frac{\partial}{\partial r}\left(r^2 \frac{\partial \Psi}{\partial r}\right) = 4\pi G \rho \tag{A.25}$$

and

$$\frac{\partial E}{\partial t} + \frac{1}{r^2}\frac{\partial}{\partial r}\left[r^2(E+p)u^r\right] = -\rho \dot{Q}_{\text{cool}} + \rho \frac{\partial \Psi}{\partial t}. \tag{A.26}$$

A.4 Fluid equations in cylindrical polar coordinates

The equations in cylindrical polar coordinates (R, θ, z) are somewhat less complex in form than the spherical polar equations. The metric is

$$ds^2 = dR^2 + R^2 \, d\theta^2 + dz^2. \tag{A.27}$$

The equations may be derived in a similar way to the spherical polar case above. The conservation of matter equation is

$$\frac{\partial \rho}{\partial t} + \frac{1}{R}\frac{\partial}{\partial R}(R\rho u^R) + \frac{1}{R}\frac{\partial}{\partial \theta}(\rho u^\theta) + \frac{\partial}{\partial z}(\rho u^z) = 0. \tag{A.28}$$

For cylindrical symmetry all $\frac{\partial}{\partial \theta}$ terms are zero, so this becomes

$$\frac{\partial \rho}{\partial t} + \frac{1}{R}\frac{\partial}{\partial R}(R\rho u^R) + \frac{\partial}{\partial z}(\rho u^z) = 0. \tag{A.29}$$

The momentum equation becomes

$$\frac{\partial u^R}{\partial t} + u^R \frac{\partial u^R}{\partial R} + \frac{u^\theta}{R}\frac{\partial u^R}{\partial \theta} - \frac{1}{R}(u^\theta)^2 + u^z \frac{\partial u^R}{\partial z} = -\frac{1}{\rho}\frac{\partial p}{\partial R} - \frac{\partial \Psi}{\partial R}, \tag{A.30}$$

$$\frac{\partial (u^\theta / R)}{\partial t} + \frac{u^R}{R}\frac{\partial u^\theta}{\partial R} + \frac{u^\theta}{R^2}\frac{\partial u^\theta}{\partial \theta} + \frac{u^\theta u^R}{R^2} + \frac{u^z}{R}\frac{\partial u^\theta}{\partial z} = -\frac{1}{\rho}\frac{\partial p}{\partial \theta} - \frac{\partial \Psi}{\partial \theta}, \tag{A.31}$$

$$\frac{\partial u^z}{\partial t} + u^R \frac{\partial u^z}{\partial R} + \frac{u^\theta}{R}\frac{\partial u^z}{\partial \theta} + u^z \frac{\partial u^z}{\partial z} = -\frac{1}{\rho}\frac{\partial p}{\partial z} - \frac{\partial \Psi}{\partial z}. \tag{A.32}$$

For cylindrical symmetry we are left with

$$\frac{\partial u^R}{\partial t} + u^R \frac{\partial u^R}{\partial R} - \frac{1}{R}(u^\theta)^2 + u^z \frac{\partial u^R}{\partial z} = -\frac{1}{\rho}\frac{\partial p}{\partial R} - \frac{\partial \Psi}{\partial R}, \tag{A.33}$$

$$\frac{\partial(u^\theta/R)}{\partial t} + \frac{u^R}{R}\frac{\partial u^\theta}{\partial R} + \frac{u^\theta u^R}{R^2} + \frac{u^z}{R}\frac{\partial u^\theta}{\partial z} = 0, \tag{A.34}$$

$$\frac{\partial u^z}{\partial t} + u^R\frac{\partial u^z}{\partial R} + u^z\frac{\partial u^z}{\partial z} = -\frac{1}{\rho}\frac{\partial p}{\partial z} - \frac{\partial \Psi}{\partial z}, \tag{A.35}$$

where we have not set $u^\theta = 0$ since cylindrical symmetry can be maintained if u^θ is independent of θ. We also have in mind that cases where these equations are useful are usually those for rotating systems.

Poisson's equation with the symmetry condition is

$$\frac{1}{R}\frac{\partial}{\partial R}\left(R\frac{\partial \Psi}{\partial R}\right) + \frac{\partial^2 \Psi}{\partial z^2} = 4\pi G\rho. \tag{A.36}$$

The energy equation becomes, again setting $\frac{\partial}{\partial \theta}$ terms to zero,

$$\frac{\partial E}{\partial t} + \frac{1}{R}\frac{\partial}{\partial R}\left[R(E+p)u^R\right] + \frac{\partial}{\partial z}\left[(E+p)u^z\right] = -\rho\dot{Q}_{\text{cool}} + \rho\frac{\partial \Psi}{\partial t}. \tag{A.37}$$

We can simplify these further with further assumptions depending on the nature of the problem, such as $u^z = 0$, or a steady state so no time dependence so all $\frac{\partial}{\partial t}$ terms become zero.

Exercises

These exercises are a mix of short questions, designed to reinforce concepts developed in the text, and longer ones, some from Cambridge examination papers, which test the reader's knowledge of the subject. The order in which they appear follows the development in the text, and not the degree of difficulty of the questions.

Chapters 1 and 2

1 Determine the equation of a general streamline of the flow $u_\phi = a$, $u_R = b$, $u_z = 0$ in cylindrical polar coordinates, and sketch the flow. Repeat for the flow $u_\phi = aR^2$, $u_R = bR^2$, $u_z = 0$. If the flows are steady, and the density at a given radius is independent of ϕ, find the radial dependence of the density in both cases.

2 Show that for a steady flow with $\nabla \cdot \mathbf{u} = 0$, the density ρ is constant along the streamlines. Need ρ be constant throughout the medium?

3 Define streamlines and particle paths and write down the differential equations for each of these.

   A flow is described by $\mathbf{u} = (u_x, u_y e^{-t/\tau}, 0)$ where u_x, u_y and τ are constants. Determine the streamlines and particle paths for such a flow and show that they are not the same at all times.

4 If $\mathbf{R} = (x, y, 0)$ and $\hat{\mathbf{R}} = \frac{\mathbf{R}}{|\mathbf{R}|}$ and a flow velocity is given for $R \geq a$ by $\mathbf{u} = U\hat{\mathbf{x}}(1 + \frac{a^2}{R^2}) - 2Ua^2 x R^{-3}\hat{\mathbf{R}}$ (where $\hat{\mathbf{x}} = \frac{\mathbf{x}}{|\mathbf{x}|}$) show that the streamlines obey $U(R - \frac{a^2}{R})\sin\phi = $ constant (where $\phi = \tan^{-1}\frac{y}{x}$). Sketch the streamlines and explain what the flow represents physically.

5 A steady 2D flow is described by $u_x = \frac{2}{x}$, $u_y = 1$. Find and sketch the streamlines. Find also a general expression for the surface density of the flow $\Sigma(x, y)$ assuming it can be written as a separable function of x and y. Radioactive nuclei are introduced in a small patch at (x_0, y_0) so as to maintain a fixed concentration there. These nuclei decay such that their number per unit mass is given by $Q = Q_0 e^{-t}$ where t is the time since introduction into the flow. Show that the surface density of radioactive nuclei (i.e. number per unit area) attains a maximum

along the radioactive streakline if x_0 is less than a critical value, and determine the coordinates of this maximum.

6 A fluid of density $\rho(x, t)$ flows in one dimension with a velocity $u(\rho) > 0$ which depends only on the density. Show that

$$\frac{\partial \rho}{\partial t} + c(\rho) \frac{\partial \rho}{\partial x} = 0,$$

where $c(\rho) = \mathrm{d}(\rho u)/\mathrm{d}\rho$.

Show that this implies that $\rho(x, t)$ is constant along a line in the (x, t) plane with slope $\mathrm{d}x/\mathrm{d}t = c(\rho)$.

At $t = 0$,

$$\rho(x, 0) = \begin{cases} \rho_0, & x < -a, \\ \rho_0 + \rho_1 \left(1 - x^2/a^2\right), & -a < x < a, \\ \rho_0, & x > a. \end{cases}$$

Sketch the approximate behaviour of density $\rho(x, t)$ at subsequent times for the cases

(a) $c = $ constant,

and

(b) $\mathrm{d}c/\mathrm{d}\rho > 0$.

If you have read Chapter 6, comment on the relevance of this to the formation of shock waves.

7 Use the summation convention to prove:

(a) $\mathbf{b} \wedge (\nabla \wedge \mathbf{b}) \equiv \nabla \left(\frac{1}{2} \mathbf{b} \cdot \mathbf{b}\right) - \mathbf{b} \cdot \nabla \mathbf{b}$,

(b) $\nabla \wedge (\nabla a) \equiv 0$,

(c) $\nabla \wedge (a\mathbf{b}) \equiv a\nabla \wedge \mathbf{b} - \mathbf{b} \wedge \nabla a$.

Using the above identities and the momentum equation, show that if $\nabla \wedge \mathbf{u} = 0$ everywhere at time $t = t_0$, then it remains so provided that the pressure is a function of the density only.

Chapter 3

8 A static infinite slab of incompressible self-gravitating fluid of density ρ occupies the region $|z| < a$. Find the gravitational field everywhere and the pressure distribution within the slab.

If a galactic disc is approximated by a uniform density slab with density $10^{-18}\,\mathrm{kg\,m}^{-3}$ and $a = 10^{18}\,\mathrm{m}$, determine the velocity of a star at the mid-plane if it starts from rest at $z = a$, and the period of its oscillation.

9 A particle is released at rest at radius R_0 from the centre of a body mass M. Compute

(a) its initial acceleration,
(b) the time it takes to reach the centre of the body

for the two cases

(i) that the body is a point mass,
(ii) that the body is a uniform sphere radius R_0.

A cluster consists initially of stars at rest, distributed in a uniform sphere. Find how long it takes a star to reach the centre as a function of its initial radius in the cluster and comment on your results.

Chapters 4 and 5

10 (a) If the Earth's atmosphere can be approximated by a perfect static gas at constant temperature of 300 K subject to a uniform gravitational field, find the variation in number density (molecules per cubic metre) with height above the Earth's surface. If the number density of molecules at the Earth's surface is $3 \times 10^{25} \, \text{m}^{-3}$, estimate the height above the Earth where the fluid approximation breaks down, and compare it with the height at which the assumption of constant gravity breaks down.

(b) The Earth runs into a cloud which is stationary with respect to the Sun. Estimate the number density the cloud would have to have in order that it seriously disturbed the Earth's atmosphere.

(Some data possibly relevant to this question: $T \sim 300 \, \text{K}$, $\mu \sim 30$, $\mathcal{R}_* = 8300$, $g = 10 \, \text{m s}^{-2}$, $M_\odot = 2 \times 10^{30} \, \text{kg}$.)

11 Estimate the temperature in the core of the Sun if the Sun is supported by gas pressure. Assume all quantities vary over a radial scale length of order the radius of the Sun. Estimate the corresponding temperature if the Sun is radiation pressure supported. Which is likely to be the case? (Radiation pressure $= \frac{1}{3}aT^4$ where $a = 7.6 \times 10^{-14} \, \text{J m}^{-3} \, \text{K}^{-4}$. The radius of the Sun is $6 \times 10^8 \, \text{m}$. $M_\odot = 2 \times 10^{30} \, \text{kg}$.)

12 The structure of a star in hydrostatic equilibrium is described by

$$M(r) = \frac{M_0}{\left[1 + \left(\frac{r_0}{r}\right)^2\right]^{3/2}},$$

where $M(r)$ is the mass enclosed within radius r, and M_0 and r_0 are constants. Determine (a) the density distribution and (b) the pressure distribution within the star.

Show that the star can be described as a polytrope, and determine the polytropic index, n.

13 A planet is composed of a material that is incompressible, density ρ, at pressures $\leq p_0$. Show that the maximum mass of such a planet that is incompressible throughout its interior is given by

$$M_{\text{max}} = \frac{2}{3\rho^2} \sqrt{\frac{1}{2\pi} \left(\frac{3p_0}{G}\right)^3}.$$

14 An equilibrium ring of isothermal fluid orbits a star at radius R. In the plane of the ring, mechanical equilibrium results from a balance of centrifugal force and the gravitational force of the central object; normal to the ring (i.e. vertically) equilibrium is between the vertical component of the gravitational force of the central object and vertical pressure gradients in the ring gas. Show that in the limit that the ring thickness $H \ll R$, the vertical density stratification in the ring is a Gaussian and determine its e-folding length in terms of the gas temperature and the angular velocity at the ring, Ω. Hence determine an upper limit to the temperature such that the ring is thin ($H \ll R$) and calculate this temperature if the ring's radius is that of the Earth's orbit around the Sun.

15 Show that if

$$\psi = -\frac{GM_s}{(r^2 + b^2)^{\frac{1}{2}}}$$

is the gravitational potential for a spherical distribution of matter then its density $\rho \propto \psi^5$.

Deduce the pressure and hence show that the equation of state is barotropic with $n = 5$.

Find the total internal energy U as a function of K, M_s and b where $K = \frac{P}{\rho^{6/5}}$.

16 Sketch the density distribution for an isothermal slab and discuss the asymptotic limits $z \to 0$, $z \to \infty$.

A galactic disc can be well approximated in its vertical structure by an isothermal slab of gas, temperature T, central density ρ_0. If a star falls from rest from a height z_0, show that its vertical velocity at height z is given by

$$\dot{z}^2 = \frac{4R_*T}{\mu} \ln\left(\frac{\cosh(az_0)}{\cosh(az)}\right).$$

17 Explain (for the case of a polytrope, index n) why the internal energy per kg, $\mathcal{E}$, is equal to $\int_0^P \frac{P}{\rho^2} \, d\rho'$, if and only if $\gamma = 1 + \frac{1}{n}$.

Calculate how the total internal energy of a polytropic star varies with stellar mass.

18 Derive the mass–radius relation for polytropic stars (equation of state $p = K\rho^{1+\frac{1}{n}}$) on the assumption that K varies with stellar mass in such a way as to maintain a constant central temperature independent of mass.

19 Show that the polytropic equation $\nabla^2 \theta = -\theta^n$ has power law solutions $\theta \alpha r^{-\zeta}$ provided $\zeta = \frac{2}{n-1}$. Give the corresponding density distribution and show that in these solutions the mass at small r is small provided $n > 3$ while the total mass at large r is small provided $1 < n < 3$.

For what values of n and ζ is the contribution to the total gravitational potential binding energy $|V|$ from material close to the origin small and when is it large?

$$\left[|V| = G \int \frac{m\,dm}{r}. \right]$$

Chapter 6

20 Show that for a linear sound wave (i.e. one in which $\Delta\rho/\rho$ is small) the velocity of fluid motion is $\ll c_s$. Estimate the maximum longitudinal fluid velocity in the case of a sound wave in air at s.t.p. in the case of a disturbance which sets up pressure fluctuations of order 0.1%.

21 A static plane-parallel atmosphere, infinite in horizontal extent and under constant gravitational acceleration $\mathbf{g}$, suffers infinitesimal adiabatic perturbations. Obtain the linearized perturbation equations:

$$\rho \frac{D\mathbf{u}}{Dt} = -\nabla p' + \mathbf{g}\rho',$$

$$\frac{D\rho'}{Dt} + \rho \operatorname{div}\mathbf{u} = 0,$$

$$\frac{Dp'}{Dt} + \gamma p \operatorname{div}\mathbf{u} = 0,$$

where p' and ρ' are the Eulerian perturbations to pressure and density.

Assuming $p' = \Re(\tilde{p}(\mathbf{x})e^{\eta t})$, $\rho' = \Re(\tilde{\rho}(\mathbf{x})e^{\eta t})$ and $\mathbf{u} = \Re(\mathbf{v}(\mathbf{x})e^{\eta t})$, where $\Re$ means real part and in which η may be complex, derive the equation

$$\eta \int \rho\mathbf{v} \cdot \mathbf{v}^* dV = \int (\tilde{p} \operatorname{div}\mathbf{v}^* + \mathbf{g} \cdot \mathbf{v}^* \tilde{\rho}) dV,$$

where the asterisk denotes complex conjugate and the integration is over the volume of the atmosphere. What assumptions, if any, did you make?

22 An isentropic atmosphere, with $p_0 = A\rho_0^{1+\frac{1}{n}}$ and A constant, in a uniform gravitational field g, has unperturbed density $\rho_0 = (kx)^n$ where x is the distance measured downwards from the top of the atmosphere, and k is a constant. Show that $g = Ak(1+n)$.

The atmosphere is subject to a small upward vertical velocity perturbation u, and accompanying adiabatic perturbations Δp $(\ll p_0)$ to the pressure and $\Delta\rho$ $(\ll \rho_0)$ to the density. Show that $\Delta p = (gx/n)\Delta\rho$.

Show also that

$$\frac{\partial}{\partial t}(\Delta\rho) - u\frac{\partial\rho_0}{\partial x} - \rho_0\frac{\partial u}{\partial x} = 0,$$

and

$$\frac{\partial u}{\partial t} + \left(\frac{\Delta\rho}{\rho_0}\right)g - \frac{g}{n\rho_0}\frac{\partial}{\partial x}(x\Delta\rho) = 0.$$

Deduce that when $n = \frac{1}{2}$,

$$\frac{\partial^2 u}{\partial t^2} = 2gx\frac{\partial^2 u}{\partial x^2} + 3g\frac{\partial u}{\partial x}. \tag{1}$$

By setting $u(x, t) = u_1(x)\exp(i\omega t)$, and $\tau = (2x/g)^{1/2}$, equation (1) may be written in the form

$$\frac{1}{\tau}\frac{d^2}{d\tau^2}(\tau u_1) + \omega^2 u_1 = 0.$$

Find the solution $u(x, t)$ which is finite at $x = 0$, and give a brief physical interpretation.

23 A large static uniform optically thin gas cloud of density ρ is radiating heat slowly into the Universe at a rate $L(p, T)$ per unit mass, where p and T are pressure and temperature. Show that if the fluid moves with velocity $\mathbf{u}$, T satisfies the equation

$$C_V\frac{dT}{dt} + \frac{p}{\rho}\operatorname{div}\mathbf{u} = -L,$$

where C_V is the specific heat capacity at constant volume, and ρ is density.

Consider a perturbation p', ρ' to p, ρ associated with an infinitesimal displacement ξ from equilibrium. Obtain the linearized equation

$$\rho\frac{\partial^2\xi}{\partial t^2} = -\nabla p' - \rho\nabla\Phi',$$

where Φ' is the perturbed gravitational potential and t is time, and write down the perturbed continuity equation. Hence show that

$$\nabla^2 p' = \frac{\partial^2}{\partial t^2}\rho' - 4\pi G\rho\rho'.$$

Chapters 7 and 8

24 Two identical clouds, radius 3×10^{16} m, temperature 10 K, collide with each other with relative velocity 4 km s^{-1}. What is the time interval, t_{coll}, over which each cloud falls into the shock? If the cooling rate in the shocked gas is $\mathcal{Q} = 10^{-10}$ J s^{-1} kg^{-1} decide whether the shock is approximately adiabatic or isothermal.

   If the clouds colliding produces an isothermal shock, what is the thickness of the shocked layer, x, at the moment that the entirety of each cloud has been shocked? At later times the layer relaxes into a structure that can be approximated by a hydrostatic isothermal slab, column density 0.1 kg m^{-2}. What fraction of the cloud masses remains within thickness x in this hydrostatic structure? (Ignore edge effects and variations of column density in the plane of the slab.)

25 Show that for a strong shock (where the upstream Mach number M_1 is large), the downstream Mach number M_2 satisfies

$$M_2^2 \simeq \frac{\gamma - 1}{2\gamma}.$$

   Hence obtain an equation for the sound-speed ratio c_2/c_1.

   A shock from a supernova travelling through the surrounding interstellar medium is observed to be travelling with speed 3000 km s^{-1}. What is the temperature immediately behind the shock? (You may assume, if you wish, the surrounding interstellar medium to have temperature 100 K and density 10^7 particles m^{-3}.)

26 Show that the momentum P per unit area of fluid between x_1 and x_2 satisfies

$$\frac{dP}{dt} = -\left[p + \rho u^2\right]_{x_1}^{x_2} + F,$$

where F is the external force per unit area on the fluid.

   The fluid encounters a stationary shock at $x = a$. Show that

$$p_2 - p_1 = (\rho_1/\rho_2)(\rho_2 - \rho_1)u_1^2,$$

where the subscripts 1 and 2 denote the values immediately upstream and downstream of the shock. Hence show that

$$\frac{p_2}{p_1} = 1 + \gamma \left(1 - \frac{\rho_1}{\rho_2}\right) M_1^2.$$

Here $M_1 = \frac{u_1}{c_1}$ is the upstream Mach number, where c_1 is the adiabatic sound speed and γ the adiabatic index.

Estimate the position of the shock in the solar wind given that the Sun loses $10^{-13} M_\odot$ per year, that the velocity immediately upstream of the shock is the escape velocity from the Sun, and that the density and temperature of the interstellar medium in the solar neighbourhood are 10^5 atoms m^{-3} and 10^4 K respectively.

27 A shocked gas is well described by the adiabatic jump conditions at the shock face, but gradually cools, becoming denser, downstream of the shock. Show that in the case of a strong shock (pre-shock pressure $p_1 \ll p_2$, the post-shock pressure) that the ratio of ram pressure to thermal pressure is always $\leq \frac{1}{2}(\gamma - 1)$ where γ is the ratio of specific heats. Hence show that, for a monatomic gas which can cool back only to the original (unshocked) temperature, the thermal pressure in the shocked gas varies by no more than 33% as the gas cools.

A crude model for the structure of shocked gas as it cools employs the above result in order to approximate the gas as being at constant thermal pressure, so that the thermal equation may be written in the form

$$c_p \frac{dT}{dt} = -Q,$$

where c_p is the specific heat at constant pressure, T the temperature, t the time, and Q the cooling rate per unit mass. If $Q = KT^2$, where K is a function of the pressure only, determine $T(t)$ (where $T(0) = T_2$, the temperature just behind the shock). Show that in this model the velocity, u, in the shocked gas satisfies $u = u_2 T(t)/T_2$ where u_2 is the velocity just behind the shock. Hence, or otherwise, show that the variation of temperature in the shocked gas with distance, x, from the shock front is given by

$$T = T_2 \exp\left(-\frac{xKT_2}{c_p u_2}\right).$$

28 Clusters of ionising stars, total luminosity L, sweep out cavities in the interstellar medium whose undisturbed density is ρ_0. Use the similarity solution method to determine the evolution of cavity radius r with time t, i.e. determine a, b, c, where $r \propto L^a \rho_0^b t^c$.

These bubbles stall when their expansion velocities become of order the sound speed in the interstellar medium. Show that the area occupied by a stalled bubble is proportional to L and hence comment on how the porosity of the interstellar medium (defined as the fractional area of the galaxy as seen from above the disc occupied by stalled bubbles) depends on how ionising stars (with given total luminosity) are organised into clusters.

If the disc of a galaxy can be approximated by a uniform density gas slab with a sharp edge at height z, comment on the different behaviours of clusters of small and large L.

Chapter 9

29 A stellar wind is maintained at a temperature of $T = 2 \times 10^6\,\text{K}$ by magnetic heating; calculate the radius at which it achieves the isothermal sound speed if the star from which it blows has mass M. You may assume the gas is atomic hydrogen. Evaluate your answer when M is the mass of the Sun $M_\odot = 2 \times 10^{30}\,\text{kg}$.

30 Isothermal gas of pressure $\rho_0 c_I^2$ and density ρ_0 at large distances from a star is steadily accreted by a star of mass M at the origin.

Calculate the accretion rate assuming that the gas remains isothermal. At what radius does the infalling gas achieve the sound velocity?

If $c_I = 1\,\text{km\,s}^{-1}$ and $n_\infty = 10^9$ hydrogen molecules m^{-3} and the mass of the hydrogen atom is $1.66 \times 10^{-27}\,\text{kg}$, evaluate this radius in terms of the solar radius $R_\odot = 7 \times 10^5\,\text{km}$.

Determine how long it will take a star of mass M to double its mass.

31 The vorticity, $\boldsymbol{\omega}$, of a fluid is defined as $\boldsymbol{\omega} = \nabla \wedge \mathbf{u}$. Find the vorticity of a fluid which is rotating with uniform constant angular velocity Ω.

Show that for an inviscid, barotropic fluid moving in a conservative force field, the vorticity satisfies

$$\frac{\partial \boldsymbol{\omega}}{\partial t} = \nabla \wedge (\mathbf{u} \wedge \boldsymbol{\omega}).$$

By evaluating $\frac{\partial}{\partial t}(\rho \boldsymbol{\omega})$, deduce that

$$\frac{D}{Dt}(\rho \boldsymbol{\omega}) = (\boldsymbol{\omega}/\rho) \cdot \nabla \mathbf{u},$$

and comment on the physical meaning of this equation.

32 Gas flows steadily and adiabatically along a smooth pipe with variable cross-sectional area $S(x)$, where x is distance measured

along the axis of the pipe. The cross-sectional area varies slowly so that the flow velocity (in the x direction) along the pipe $v(x)$ and the density $\rho(x)$ may be treated as functions of x alone. Thus the rate of mass flow along the pipe, $\dot{M}$, may be written as

$$\dot{M} = \rho v S = \text{constant}. \tag{1}$$

At $x = 0$, $S(x)$ is sufficiently large that the flow is very subsonic. If the gas has ratio of specific heats $\gamma = 2$, show that approximately

$$\frac{1}{2}v^2 + c^2 = c_0^2, \tag{2}$$

where c is the speed of sound in the gas, and c_0 is its value at $x = 0$.

Sketch the relationship (2) in the (v, c) plane.

As x increases, $S(x)$ decreases monotonically to a minimum value $S_{\min}$ at $x = x_{\min}$ and then increases monotonically thereafter. In the same (v, c) plane plot the relationship (1) for various values of $S(x)$, taking care to mark the curve corresponding to $S_{\min}$.

Show that at $x_{\min}$, $v = \sqrt{\frac{2}{3}}c_0$.

If $S \to \infty$ as $x \to \infty$, find v/c_0 as $x \to \infty$.

33 A narrow, highly supersonic jet is emitted into a vacuum in the $z = 0$ plane in (R, ϕ, z) coordinates.

(a) The jet is emitted at constant speed V_R ($\gg c_s$) but with variable direction such that the initial velocity is in the direction

$$\phi(t) = \beta \sin\left(\frac{2\pi t}{P}\right),$$

where β and P are constants and $\beta = 45°$. Sketch a snapshot of the flow, indicating characteristic dimensions of the flow and the velocity vectors of the fluid elements.

(b) The jet is emitted with constant direction but variable speed

$$V_R(t) = V_0 + V_1 \sin\left(\frac{2\pi t}{P}\right),$$

where V_0 and V_1 are constants and $V_0 \gg V_1 \gg c_s$. At what distance from the origin can pressure no longer be ignored?

34 A jet propagates in the z direction through a hydrostatic slab of isothermal gas, temperature T_s (i.e. one with density distribution $\rho = \rho_0 \text{sech}^2\left(\frac{z}{z_s}\right)$). If the jet is also isothermal, temperature T_j, write down the density distribution in the jet, $\rho_j(z)$, explaining carefully what boundary condition applies at the jet/slab interface.

Show that the minimum cross-sectional area of the jet, $A_{\min}$, is given by

$$A_{\min} = \frac{\dot{M}}{\rho_{0j} c_j} \exp\left[\frac{\mu}{2 R_* T_j} \left(c_j^2 - \left(\frac{\dot{M}}{\rho_{0j} A_0} \right)^2 \right) \right],$$

where $\dot{M}$ is the mass flux in the jet, c_j is the sound speed in the jet and ρ_{0j}, A_0 are the density and cross-sectional area at the base of the jet $(z = 0)$.

Write down the height, $z_{\min}$, of this minimum and the jet velocity at this point. Show furthermore that the cross-sectional area as a function of z, $A(z)$, is

$$A(z) = \frac{A_0 \cosh^2\left(\dfrac{z}{z_s} \right)}{\left[1 + 2\left(\dfrac{A_0 \rho_{0j}}{\dot{M}} \right)^2 \dfrac{R_* T_j}{\mu} \ln\left[\cosh^2\left(\dfrac{z}{z_s} \right) \right] \right]^{1/2}}$$

and sketch the shape of the jet. (Ignore the effects of gravity on the jet structure.)

Chapter 10

35 A star is described by a polytropic equation of state, index n. For what values of n is such a star stable against convection? In the case that the star is convectively unstable, describe its structure and how energy is transported from its core to a distant point outside the star. (Assume that the star consists of fully ionised monatomic hydrogen.)

36 A self-gravitating slab of gas is heated by cosmic rays (constant heating rate per unit mass) and cooled by optically thin thermal bremsstrahlung for which the cooling rate per unit mass is proportional to $\rho T^{0.5}$. Discuss the stability of the slab to (a) thermal and (b) Rayleigh–Taylor instabilities.

37 A medium is in thermal balance between cosmic ray heating with constant heating rate H per unit mass and optically thin bremsstrahlung cooling, for which the cooling rate per unit mass is proportional to $\rho\sqrt{T}$, where ρ is the density and T is the temperature. Show that for any flow for which this balance is maintained the gas pressure p satisfies

$$p = A\rho^{-1},$$

where A is a constant.

Determine whether or not a uniform medium with such a pressure–density law is stable to small perturbations.

Comment on the physical interpretation of this result.

38 A perfect gas is in hydrostatic equilibrium in a uniform gravitational field. Show that the configuration is stable against adiabatic convection if the temperature gradient dT/dz is such that

$$\left|\frac{dT}{dz}\right| < \left(1 - \frac{1}{\gamma}\right)\frac{T}{p}\left|\frac{dp}{dz}\right|,$$

where p is the gas pressure and γ the ratio of specific heats.

A self-gravitating static slab of gas has a polytropic index $n = 1$, so throughout its structure $p \propto \rho^2$. Determine the density and temperature structure of the slab and show that it is convectively unstable.

State briefly what will happen to the pressure–density relation in the slab as a result of this instability.

39 Calculate the ratio of the free-fall time to the sound-wave-crossing time for a uniform gaseous sphere containing one Jeans mass.

Show that if such a sphere contracts homogeneously (i.e. maintaining uniform density) and isothermally, the number of Jeans masses it contains rises as $\mathfrak{R}^{3/2}$ where $\mathfrak{R}$ is the collapse factor (i.e. the ratio of the radius to its initial value).

40 A thin disc of gas, rotating around an object mass M, is supported in the vertical (z) direction by a balance between pressure forces and the z component of the central object's mass. Write down the equation of hydrostatic equilibrium in the z direction. If the density distribution is of the form $\rho(z) \propto (z_m^2 - z^2)^2$ deduce the pressure distribution and polytropic index, n, for the disc. Is the disc stable against convection if composed of (a) monatomic, (b) diatomic gas?

41 (a) A smooth pipe, with square cross-section of side a, lies flat on a horizontal surface, in a uniform vertical gravitational field, g. The pipe contains incompressible fluid. The fluid in the upper half of the pipe has density ρ_1, and the fluid in the lower half of the pipe has density ρ_0, where $\rho_1 > \rho_0$. Name the instability which ensues and identify the energy source which powers it.

Calculate the energy E_1 available to the instability per unit length of the pipe.

(b) Now consider the case $\rho_1 < \rho_0$ but with the fluid in the upper half pipe moving along the pipe with velocity V and with the fluid in the lower half pipe moving along the pipe with velocity $-V$.

Compare the quantities E_0 (from Part (a)) and E_1 to explain why the size of the Richardson number,

$$Ri = \frac{\rho_0 \rho_1}{\rho_0^2 - \rho_1^2} \frac{V^2}{ag},$$

is relevant to stability of the flow.

42 The heat loss rate function $Q(p, T)$ per unit mass is given by

$$Q = \frac{1}{4} \alpha p \left[(\theta - 3)^3 - 3(\theta - 3) + 18 \right] - \gamma,$$

where $\theta = T/T_c$ and α, γ and T_c are positive constants. Draw a rough sketch of $\dot{Q}$ as a function of θ. Find the turning point of $\dot{Q}$ as a function of temperature for fixed pressure and the values of $\dot{Q}$ there. Hence show that there are three steady states if the pressure lies between $\gamma/4\alpha$ and $\gamma/5\alpha$. Which of these steady states are thermally unstable? Show that if the pressure is slowly increased from below $\gamma/5\alpha$ to above $\gamma/4\alpha$ and then slowly decreased again the system does not return by the same path in (p, T) space.

Chapters 11 and 12

43 An incompressible fluid of density ρ with constant viscosity coefficient η flows along an annular pipe of length ℓ in the region between the inner radius R_1 and the outer radius R_2. Determine the mass flow rate Q through the pipe if the pressure at one end of the pipe is p_1 and the other end it is p_2.

44 A layer of incompressible fluid of thickness h is bounded above by a free surface and below by a fixed plane inclined at an angle α to the horizontal in a uniform gravitational field with gravitational acceleration g. Show that the flow rate (per unit length perpendicular to the flow) due to gravity is $Q = \rho g h^3 \sin \alpha / 3\nu$, where ν is the kinematic viscosity and ρ the fluid density.

45 Suppose there is a unidirectional flow $u_x(y, t = 0)$ in an infinite viscous fluid at time $t = 0$. Show that the flow remains unidirectional, and evolves with time as

$$u_x(y, t) = \frac{1}{2\sqrt{\pi \nu t}} \int_{-\infty}^{\infty} u_x(y', 0) \exp \left[-\frac{(y - y')^2}{4\nu t} \right] dy'$$

if there is no pressure gradient.

46 Consider a static fluid with uniform density ρ_0 and pressure p_0. Assume that the fluid obeys the equations

$$\frac{\partial \rho}{\partial t} + \nabla \cdot (\rho \mathbf{v}) = 0,$$

$$\frac{\partial \mathbf{v}}{\partial t} + (\mathbf{v} \cdot \nabla) \mathbf{v} = -\frac{1}{\rho} \nabla p - \frac{\eta}{\rho} \left[\nabla^2 \mathbf{v} + \frac{1}{3} \nabla (\nabla \cdot \mathbf{v}) \right] - \frac{\zeta}{\rho} \nabla (\nabla \cdot \mathbf{v}),$$

where η and ζ are the coefficients of shear and bulk viscosity, which may be assumed constant. Show that the fluid supports small amplitude plane-wave perturbations of the form

$$\delta v \approx A e^{-\gamma x} e^{i(kx - \omega t)},$$

$$\delta \rho = \rho - \rho_0 \approx B e^{-\gamma x} e^{i(kx - \omega t)},$$

where $\gamma \ll k$ and

$$\gamma \approx \frac{\omega^2}{2\rho_0 c_s^3} \left(\frac{4}{3}\eta + \zeta \right), \qquad c_s \approx \frac{\omega}{k},$$

and $c_s^2 = (dp/d\rho)$ is the sound speed.

Give a physical explanation of this result.

47 Show that if a gas in an accretion disc is in a state of hydrostatic equilibrium between the vertical component of the gravitational acceleration of the central object and the vertical pressure gradient, then the angular frequency of rotation, Ω, the sound speed, c_s, and the vertical scale height of the disc, H, are related by $c_s \sim H\Omega$. Show also that if the $r\phi$ component of the viscous stress tensor is equal to some factor α times the thermal pressure then (i) $\nu \sim \alpha c_s H$ and (ii) the Reynolds number $\mathcal{R} \sim \frac{1}{\alpha} \left(\frac{r}{H} \right)^2$.

48 Write down an expression for the monochromatic flux, F_ν, at frequency ν emitted by an optically thick steady state Keplerian accretion disc. Show that for a disc where the outer radius is much greater than the inner one, to a good approximation $F_\nu \propto \nu^{\frac{1}{3}}$ over some frequency range. Explain for what range of frequencies this form is valid.

49 Show that the viscous evolution equation for a Keplerian accretion disc

$$\frac{\partial \Sigma}{\partial t} = \frac{3}{R} \frac{\partial}{\partial R} \left[R^{\frac{1}{2}} \frac{\partial}{\partial R} \left(\nu \Sigma R^{\frac{1}{2}} \right) \right]$$

may be written in the form

$$\frac{\partial S}{\partial t} = \frac{\partial^2 Y}{\partial X^2},$$

where $S = \Sigma R^{\frac{3}{2}}$, $Y = \nu \Sigma R^{\frac{1}{2}}$ and $X = 2R^{\frac{1}{2}}$. Derive also a partial differential equation for $S(X, t)$ in the case where $Y = Y_0 + \beta(S - S_0)$, where Y_0, S_0 and β are constants. Describe the evolution of S in the cases that (a) $\beta > 0$, and (b) $\beta < 0$.

Chapter 13

50 An infinite homogeneous stationary fluid has uniform density ρ_0 and pressure p_0, and is permeated by a constant magnetic field $\mathbf{B}_0 = (B_0, 0, 0)$ in Cartesian coordinates. A small velocity perturbation

$$\mathbf{u} = (0, u_1(y, t), 0)$$

gives rise to a perturbation of the magnetic field of the form

$$\mathbf{B}_0 = (B_0 + B_1(y, t), 0, 0)$$

with $B_1 \ll B_0$, and gives rise to adiabatic perturbations to pressure and density such that $p = p_0 + p_1(y, t)$ and $\rho = \rho_0 + \rho_1(y, t)$. Show that to linear order

$$\frac{\partial^2 u_1}{\partial t^2} = (c_{s0}^2 + v_A^2) \frac{\partial^2 u_1}{\partial y^2},$$

where c_{s0} is the sound speed and v_A the Alfvén speed in the unperturbed fluid.

What is the physical meaning of this equation?

51 A gas is predominantly supported against gravity by magnetic pressure which scales as B^2 for magnetic flux density B. By comparing the gravitational collapse timescale to the propagation timescale for magnetic disturbances, show that the magnetic Jeans mass scales as B^3/ρ^2.

Hence show that if a uniform spherical cloud with a frozen-in magnetic field contracts homogeneously, the number of magnetic Jeans masses it contains is constant.

52 An isothermal gas at temperature T is in pressure equilibrium under a uniform gravitational field g which acts in the $-z$ direction. Show that the density is given by

$$\rho = \rho_0 \exp\left(\frac{-\mu g z}{\mathcal{R}_* T}\right),$$

where ρ_0 is the density at $z = 0$, $\mathcal{R}_*$ is the gas constant and μ the mean molecular weight.

If the gas is fully ionised and contains a magnetic field $\mathbf{B} = (B(z), 0, 0)$ where $B(z) = B_0(\rho/\rho_0)^{1/2}$, what is the density as a function of the height z?

If $B(z) = B_0(\rho/\rho_0)$, what is the relationship between the density ρ and the height z?

Compare the scale heights relative to $z = 0$ for the three cases and comment on the effect that the magnetic field has on this quantity.

53 (a) Show that Bernoulli's equation applies for a steady hydromagnetic flow in which the magnetic field is everywhere parallel to the streamlines.

(b) An ideal hydromagnetic flow is such that the velocity $\mathbf{u}$ is perpendicular to the magnetic field $\mathbf{B} = (0, 0, B)$, and there is no z dependence for any of the variables. Show that the momentum equation has a similar form to the standard fluid one, and that the time evolution of the magnetic field is given by $\frac{\partial B}{\partial t} = -\nabla \cdot (B\mathbf{u})$.

54 In cylindrical polar coordinates, (R, ϕ, z), a cylindrical flux tube has a magnetic field

$$\mathbf{B} = (0, 0, B(R)).$$

Use the equation of hydrostatic equilibrium

$$\nabla p = \mathbf{j} \wedge \mathbf{B},$$

where p is the pressure, to show that $p + \frac{1}{2}B^2/\mu_0$ is independent of R.

A flux tube with $B \sim 0.1\,\mathrm{T}$ emerges through the photosphere of a magnetic K star. Does the flux tube have a significant effect on conditions at the photosphere when the temperature $T = 4000\,\mathrm{K}$ and density $\rho = 5 \times 10^{-4}\,\mathrm{kg\ m^{-3}}$?

(You may assume that for the field described here $\mathbf{J} = (0, -\frac{1}{\mu_0}\frac{dB}{dR}, 0)$.)

55 Consider a static pressure balanced magnetic field in which all variables are independent of z, i.e. $\frac{\partial}{\partial z} = 0$. If we write the magnetic field as

$$\mathbf{B} = B_z(x, y)\hat{\mathbf{e}}_z + \nabla \wedge (A(x, y)\hat{\mathbf{e}}_z)$$

(as in the case of the stream function for two-dimensional flows), show that $p + \frac{B_z^2}{2\mu_0}$ has to be a function $F(A)$ of A. Show further that

$$\frac{\partial^2 A}{\partial x^2} + \frac{\partial^2 A}{\partial y^2} + \mu_0 \frac{dF}{dA} = 0.$$

56 Consider a constant initial magnetic field $\mathbf{B} = B_0\hat{\mathbf{e}}_y$ in an ideal plasma. Suppose a velocity field $\mathbf{u} = u_0 e^{-y^2}\hat{\mathbf{e}}_x$ is switched on at time $t = 0$. Determine how the magnetic field evolves with time.

Books for background and further reading

Fluids dynamics books usually concentrate on incompressible fluids. Some which provide a good coverage are:

D. Acheson, *Elementary Fluid Dynamics*, Oxford University Press (1994).

G. K. Batchelor, *An Introduction to Fluid Dynamics*, Cambridge University Press (re-issued 2000).

H. Lamb, *Hydrodynamics*, Cambridge University Press (6th edn. 1932, reprinted 1993).

L. D. Landau & E. M. Lifshitz, *Fluid Mechanics*, Pergamon Press (1987).

M. J. Lighthill, *An Informal Introduction to Theoretical Fluid Mechanics*, Oxford University Press (1993).

In an astrophysical context the following texts may be useful, though the treatment is more advanced:

J. E. Pringle & A. R. King, *Astrophysical Flows*, Cambridge University Press (2007).

F. Shu, *The Physics of Astrophysics: Gas Dynamics*, University Science Books (1992).

For topics covered in particular chapters a partial list is provided below.

Chapter 4

G. W. Collins II, *The Fundamentals of Stellar Astrophysics*, W. H. Freeman & Co. (1989), includes details of energy transfer processes relevant for stars, a number of which have wider applicability (also referred to in Chapter 9).

D. E. Osterbrock, *Astrophysics of Gaseous Nebulae and Active Galactic Nuclei*, University Science Books (1989), provides an excellent introduction to photoionisation, and heating and cooling processes in gaseous nebulae.

A. G. G. M. Tielens, *The Physics and Chemistry of the Interstellar Medium*, Cambridge University Press (2005), gives an overview of the interstellar medium in galaxies.

Chapter 5

S. Chandrasekhar, *An Introduction to the Study of Stellar Structure*, Dover (1957), contains an exhaustive treatment of polytropic gas spheres.

J. Binney & S. Tremaine, *Galactic Dynamics*, Princeton University Press (1994), deals primarily with the dynamics of stellar systems, and so collisionless systems.

R. W. Hilditch, *An Introduction to Close Binary Stars*, Cambridge University Press (2001), gives comprehensive coverage of many aspects of binary stars.

Chapter 10

C. F. McKee, in *The Physics of the Interstellar Medium and Intergalactic Medium*, ASP Conference Series, Vol. 80, A. Ferrara, C. F. McKee, C. Heiles & P. R. Shapiro (eds.) (1995), provides a review of physics of the multi-phase interstellar medium.

Chapter 12

J. Frank, A. R. King & D. J. Raine, *Accretion Power in Astrophysics*, Cambridge University Press (1992), gives a fuller description of the physical processes in accretion discs.

J. E. Pringle, 'Accretion discs in astrophysics' in *Annual Review of Astronomy and Astrophysics*, Vol. 19, Annual Reviews Inc. (1981), pp. 137–162, gives a very good description of the basic physics.

Chapter 13

B. I. Bleaney & B. Bleaney, *Electricity and Magnetism*, Volume 1, Oxford University Press (1989), is a classic text on electricity and magnetism.

R. O. Dendy (ed.), *Plasma Physics: An Introductory Course*, Cambridge University Press (1993), gives an in-depth treatment of the basic principles of the subject.

Index

Printed in the United States
By Bookmasters